Lecture Notes in Mathematics

A collection of informal reports and seminars
Edited by A. Dold, Heidelberg and B. Eckmann, Zürich

92

Category Theory, Homology Theory and their Applications II

Proceedings of the Conference held at the Seattle Research Center of the Battelle Memorial Institute, June 24 – July 19, 1968
Volume Two

1969

Springer-Verlag Berlin · Heidelberg · New York

Library of Congress Catalog Card Number 75-75931 Printed in Germany. Title No. 3698

Preface

This is the second part of the Proceedings of the Conference on Category Theory, Homology Theory and their Applications, held at the Seattle Research Center of the Battelle Memorial Institute during the summer of 1968. The first part, comprising 12 papers, was published as Volume 86 in the Lecture Notes series.

Following the Table of Contents, there is appended a list of papers to be published in subsequent volumes.

It is again a pleasure to express to the administrative and clerical staff of the Seattle Research Center the appreciation of the contributors to this volume, and of the organizing committee of the conference, for their invaluable assistance in the preparation of the manuscripts.

Cornell University, Ithaca, January,1969 Peter Hilton

Table of Contents

Papers to appear in future volumes

J. F. Adams	Generalized cohomology
J. Beck	On H-spaces and infinite loop spaces
D. B. Epstein and M. Kneser	Functors between categories of vector spaces
D. B. Epstein	Natural vector bundles
P. J. Freyd	New concepts in category theory
J. W. Gray	Categorical fibrations and 2-categories
R. Hoobler	Non-abelian sheaf cohomology
M. Karoubi	Foncteurs dérivés et K-théorie
F. E. J. Linton	Relative functorial semantics
E. G. Manes	Minimal subalgebras for dynamical triples
P. May	Categories of spectra and infinite loop spaces
P. Olum	Homology of squares and factorization of diagrams

RELATIONS FOR GROUPS AND FOR EXACT CATEGORIES

by

Hans-Berndt Brinkmann

Relations (Correspondences) provide a useful tool for diagram chasing and for defining connecting morphisms, higher order cohomology operations or differentials in spectral sequences. They have been studied by Riguet, Lambek, MacLane, Puppe, Hilton and Leicht among others.

Puppe posed the question whether any exact[(1)] category (§1) admits a calculus of relations satisfying certain axioms and thus being suited for the purposes mentioned in the beginning (Puppe [10;4.18, p.18 and Axioms K1-3]). It was also shown in [10; 4.15, p.17] that at most one such calculus of relations can exist. We give a positive answer to the problem. The construction described below can be carried out for any category. Applied to a locally and colocally small exact category, we obtain a calculus of relations as desired. Applied to the category of groups we recover the standard category of homomorphic relations for groups. A construction of relations for abelian categories was given by Puppe [10]. Hilton [4] then described this case by means of a fractional calculus. For exact categories the problem was independently also considered by Leicht [6] and Calenko [3].

The present note is, except for some minor changes an abbreviated version of [1]. The changes made are such that the case of groups is included now.

1. A category is exact[(1)], if it has a zero object and if every morphism is a composition of a cokernel followed by a kernel. The decomposition readily turns out to be unique up to isomorphism and functorial after choice. Standard properties of exact categories may be found in Mitchell [9; I.7-15, p.9-19] or in [2].

[(1)] The former terminology is quasiexact [10]. No addition is required as in the exact categories of Buchsbaum.

Exact categories need not be abelian (Puppe [10;§7, Beispiel A, p.23]. The additive group Z of integers is a category with one object. Adjoining a zero object we obtain an exact category which does not even admit a (natural) addition, since else we would obtain a field having Z as its multiplicative group.

2. A category of relations consists of a category $\underline{K}$, a (natural) order relation $<$ on $\underline{K}$ (i.e. compatible with composition) and a contravariant order preserving functor $^{\#} : \underline{K} \longrightarrow \underline{K}$ (conversion) satisfying $A^{\#} = A$ for objects and $f^{\#\#} = f$ for morphisms. For simplicity we write $\underline{K}$ for the triple of data $(\underline{K}, <, ^{\#})$. Details may be found in [10;1.3, p.3-4].

Examples: Relations for sets $R\underline{S}$, pointed sets $R\underline{S}_*$, groups $R\underline{G}$, abelian groups $R\underline{Ab}$.

2.2. Let $\underline{K}$ be a category of relations. $f \in \underline{K}$ will be called a map, iff

		*
	--	$f^{\#}f < 1$ [2]
$<$	$ff^{\#} > 1$	-- .

The arrangement of the picture refers to the two types of duality involved in the definition. The maps of $\underline{K}$ form a subcategory $M\underline{K}$, which contains all objects of $\underline{K}$ (Identities are maps). Obviously we have

Lemma 2.3 $f < g$ and $f,g \in M\underline{K}$ imply $f = g$.

Proof: $g < ff^{\#}g < fg^{\#}g < f$ using $1 < ff^{\#}$, $f^{\#} < g^{\#}$ and $g^{\#}g < 1$.

$M\underline{K}$ hence so far carries no further structure than being a category ($^{\#}$ does not restrict to $M\underline{K}$, since for $f \in M\underline{K}$, $f^{\#}$ need not be a map).

Examples: $MR\underline{S} = \underline{S}$ (sets), $MR\underline{S}_* = \underline{S}_*$ (pointed sets), $MR\underline{G} = \underline{G}$ (groups), $MR\underline{Ab} = \underline{Ab}$ (abelian groups).

---

(2) fg means first f then g

3. Let $\underline{K}$ be a category of relations. If further axioms are imposed on $\underline{K}$, it turns out that $\underline{MK}$ is a (locally and colocally small) exact category (Axioms K1-3 of [10]; [10]; 4.8, p.15]). We will not go into enough details to bother the reader with the precise form of these axioms. Also it turns out that the embedding of $\underline{MK}$ into $\underline{K}$ is especially nice [10; 4.8-4.12, p. 15-16] and that $\underline{K}$ is up to isomorphism determined by $\underline{MK}$ [10; 4.15, p.17]. $\underline{K}$ satisfying axioms K1-3 of [10] will be called a pseudoexact category of relations.

Examples: The axioms are satisfied for $\underline{RAb}$. Except for K3b they are also satisfied for $\underline{RG}$. K3b, however, in this case may be weakened so as to still recover uniqueness (E.g. replace K3 by: Every u such that $u = u^{\#} = u^2$ (symmetric idempotent) may be expressed in the form $u = m^{\#}ee^{\#}m$ (subquotient) such that m, e are maps and $mm^{\#} = 1$ and $e^{\#}e = 1$.

4. We can now state the following result:

Theorem 4.1

Let $\underline{C}$ be a locally and colocally small exact category. Then there exists a pseudoexact category of relations $\underline{K}$ such that $\underline{C} = \underline{MK}$.

The smallness condition of the theorem is only used to insure that $\underline{K}$ has hom sets. The construction below is believed to give an honest mathematical object independent of this assumption.

Let $\underline{Ex}$ denote the (illegitimate) category of locally and colocally small exact categories and $\underline{Psex}$ the (illegitimate) category of pseudoexact categories of relations both with their appropriate functors. With regard to the results of [10] mentioned in §3, 4.1 results in that we defined a functor $\underline{Ex} \xrightarrow{K} \underline{Psex}$ such that the pair $\underline{Ex} \underset{M}{\overset{K}{\rightleftarrows}} \underline{Psex}$ is an equivalence of categories (M is the extension of "maps" to a functor).

To motivate the construction to be given, we assume that $\underline{K}$ is a pseudoexact category of relations. We consider a diagram

(4.2)

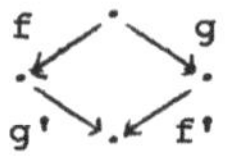

in $\underline{MK}$ and analyze when

(4.3) $$f^{\#}g < g'f'^{\#} \quad \text{and}$$

(4.4) $$f^{\#}g = g'f'^{\#} .$$

We have

Lemma 4.5 $f^{\#}g < g'f'^{\#}$, iff the diagram commutes in $\underline{MK}$ $(fg' = gf')$.

Lemma 4.6 $f^{\#}g = g'f'^{\#}$, iff the decomposition in $\underline{MK}$ of the diagram to

(4.7)

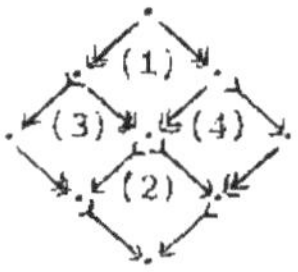

yields (1) a pushout, (2) a pullback and (3) and (4) bicartesian (pullback and pushout) in $\underline{MK}$.

A proof is given in [2], the hypotheses may be weakened to the form mentioned in §3 Examples (See [2; Anhang]).

A square as 4.2 in $\underline{MK}$ such that $f^{\#}g = g'f'^{\#}$ will be called fully commutative. If $\underline{C}$ is abelian, Hilton [4; Theorem 3.3, p.258] described fully commutative squares by the existence of a completion to a commutative diagram

(4.8)

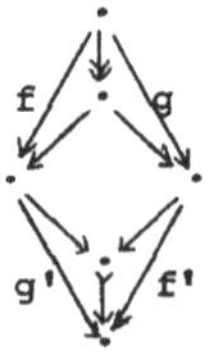

such that the inner square is bicartesian. This is readily seen to be equivalent to the characterization given by 4.6 .

5. We are thus led to the following construction: Let $\underline{C}$ be a category. For every morphism $A \xrightarrow{f} B$ in $\underline{C}$ we adjoin a symbol $f^{\#}$. The possibly illegitimate category cword$\underline{C}$ of cwords on $\underline{C}$ is then defined as follows[(3)]:

Every object A of $\underline{C}$ is a cword from A to A and it is an identity of cword$\underline{C}$,
Every map $A \xrightarrow{f} B$ is a cword from A to B,
For every map $A \xrightarrow{f} B$, $f^{\#}$ is a cword from B to A,
If α is a cword from A to B and if β is a cword from B to C and if neither α nor β is an object of $\underline{C}$, then $\alpha \otimes \beta$ is a cword from A to C.

We can hence identify cword$\underline{C}$ with the category of all finite diagrams as e.g.

(5.1) $$A \xrightarrow{f_1} \cdot \xleftarrow{f_2} \cdot \cdots \cdot \xrightarrow{f_{n-1}} \cdot \xrightarrow{f_n} B$$

having concatenation as composition. Here $\cdot \xleftarrow{f} \cdot$ abbreviates $f^{\#}$.

Duality of $\underline{C}$ will be denoted by * . We have

(5.1)* $$A \xleftarrow{f_1} \cdot \xrightarrow{f_2} \cdot \cdots \cdot \xleftarrow{f_{n-1}} \cdot \xleftarrow{f_n} B.$$

$A \longmapsto A$, $f \longmapsto f^{\#}$, $f^{\#} \longmapsto f$ extends to an antiautomorphism of cword$\underline{C}$. This will also be denoted by # . We have

(5.1)# $$B \xleftarrow{f_n} \cdot \xleftarrow{f_{n-1}} \cdot \cdots \cdot \xrightarrow{f_2} \cdot \xleftarrow{f_1} A.$$

Let R be the binary relation between cwords defined by

(5.2) $(A \xrightarrow{1_A} A, A) \in R$ for all A,

(5.3) $(f \otimes g, fg) \in R$ for all $\cdot \xrightarrow{f} \cdot \xrightarrow{g} \cdot$,

(5.4) $(m_1 \otimes m_2^{\#}, m_1'^{\#} \otimes m_2') \in R$, if 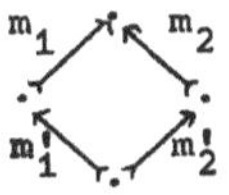 is a pullback in $\underline{C}$,

(5.5) $(m \otimes e^{\#}, e'^{\#} \otimes m') \in R$ if 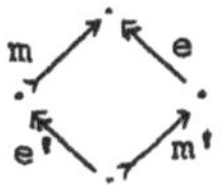 is bicartesian in $\underline{C}$.

---

(3) The terminology is due to Freyd.

Let $\sim$ be the natural equivalence relation on cword$\underline{C}$ generated by $R \cup R^* \cup R^{\#} \cup R^{\#*}$, where e.g. R^* has the obvious meaning $(\alpha,\beta) \in R^* \Longleftrightarrow (\alpha^*,\beta^*) \in R$. Obviously $\sim$ is compatible with $^{\#}$ and we obtain a quotient category $K\underline{C}$: = cword$\underline{C}/_{\sim}$ equipped with an involutive automorphism $^{\#}$, which is the identity on objects. Also there is an obvious functor $\underline{C} \longrightarrow K\underline{C}$. The image of $f \in \underline{C}$ under this functor will be denoted by f.

Theorem 5.6

Let $\underline{C}$ be a (locally and colocally small) exact category. Then a natural order relation $<$ may be introduced in $K\underline{C}$ by means of $f^{\#}g < g'f'^{\#}$ for commutative diagrams

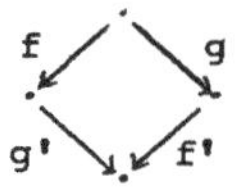

in $\underline{C}$ and furthermore $K\underline{C}$ with $^{\#}$ and $<$ is a pseudoexact category of relations such that $MK\underline{C} = \underline{C}$.

In [1] 5.6 is proved in the following form:

Theorem 5.7

Let $\underline{C}$ be a locally and colocally small exact category. Then every morphism Φ in $K\underline{C}$ has a unique representative $\cdot \xleftarrow{m} \cdot \xrightarrow{f} \cdot \xleftarrow{e} \cdot$ in cword$\underline{C}$ such that m is a subobject[(4)] and e is a quotient object[(4)] in $\underline{C}$. We introduce a relation "$<$" by $\Phi < \Phi'$, iff there is a commutative diagram

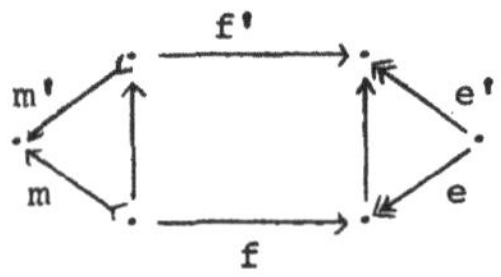

in $\underline{C}$ for the representatives. Then $K\underline{C}$ with $<$ and $^{\#}$ is a pseudoexact category of

---

(4) i.e. selected from its class

relations such that $MK\underline{C} = \underline{C}$.

The proof given in [1] uses the methods of MacLane [8][5]. The idea may be briefly described as follows: It is not hard to see that every cword is equivalent to some standard form $\cdot \longleftarrow\!\!\!< \cdot \longrightarrow \cdot \twoheadleftarrow \cdot$ as described in the theorem. For the uniqueness we analyze the process of shortening cwords to this standard form and introduce a relation "shorter" among cwords. We then show that the standard form is the shortest form possible and that any two cwords obtained by shortening the same cword admit a common shortening. To do this, we use the following

Lemma 5.8 (Proof: [2], [1]) Let $\underline{C}$ be an exact category. Let

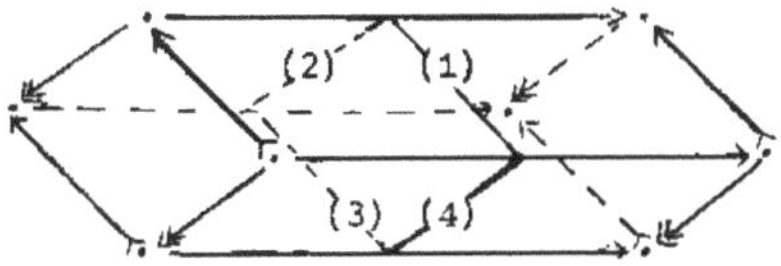

be a commutative diagram in $\underline{C}$. If (1) is a pullback and (2) is a pushout, then (3) is a pullback and (4) is a pushout.

Finally the equivalence relation generated by "shorter" is $\sim$, the relation used in defining $K\underline{C}$. This obviously suffices to prove the uniqueness (The uniqueness is used to define $<\underset{\cdot}{|}$). The rest of the proof (Axioms K1-3 etc.) is then straightforward.

The relation defined in 5.4,5.5 differs slightly from the relation used in [1]. By lemma 5.8 they agree for exact categories. The change is made to include relations for groups in the construction (see §6).

It is clear that the quotient of cword$\underline{C}$ by the natural relation generated by 5.2, 5.3 is the (illegitimate) free $^{\#}$-category on $\underline{C}$. $K\underline{C}$ then is a quotient of this free category. cword$\underline{C}$ may thus be avoided and it appears only for technical reasons pertaining to the proof of 5.7.

---

(5) That these methods could be applied here occured to me while reading MacDonald [7].

6. Starting from $m^{\#}e'm'e^{\#}$ we may form the following diagram in $\underline{C}$

(6.1)

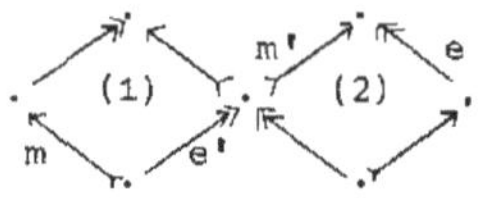

by means of a pushout (1) and a pullback (2). 5.8 then shows that both squares are bicartesian. It follows that any of the forms $\cdot \leftarrowtail \cdot \twoheadrightarrow \cdot \rightarrowtail \cdot \twoheadleftarrow \cdot$,
$\cdot \leftarrowtail \cdot \twoheadrightarrow \cdot \twoheadleftarrow \cdot \rightarrowtail \cdot$,
$\cdot \twoheadrightarrow \cdot \leftarrowtail \cdot \rightarrowtail \cdot \twoheadleftarrow \cdot$,
$\cdot \twoheadrightarrow \cdot \leftarrowtail \cdot \twoheadleftarrow \cdot \rightarrowtail \cdot$ could have been used in starting theorem 5.7.

4.5 and 4.6 are true for the category of relations for groups(6) and enough fully commutative squares exist to reduce every relation in $\underline{RG}$ to the form $\cdot \leftarrowtail \cdot \twoheadrightarrow \cdot \twoheadleftarrow \cdot \rightarrowtail \cdot$ (Lambek [5; Propositions 1,2, p. 47,48]). None of the other forms can in general be constructed for groups, however, since 5.8 fails for groups and (1) and (2) in 6.1 formed as pushout need not be bicartesian (or fully commutative): A pushout

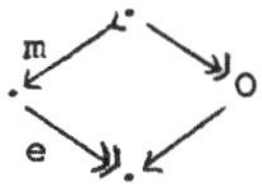

for groups, where m is not normal cannot be a pullback and hence cannot be fully commutative (pushout → e = cokm, pullback → m = kere !). $\underline{KG}$ with order relation as in 5.6 is $\underline{RG}$.

Certainly we could have defined relations for groups and exact categories directly using $\cdot \leftarrowtail \cdot \twoheadrightarrow \cdot \twoheadleftarrow \cdot \rightarrowtail \cdot$. Then the problem arises whether the composition is associative. Diagram lemmas similar to 5.8 have to be used in asserting this. $\cdot \leftarrowtail \cdot \twoheadrightarrow \cdot \twoheadleftarrow \cdot \rightarrowtail \cdot$ was (for exact categories) used by Calenko [3],

(6) 4.5 does not use axioms K1-3. As mentioned before, 4.6 may be proved under the weaker assumptions mentioned in §3 Example.

$\cdot \leftarrowtail \cdot \longrightarrow \cdot \twoheadleftarrow \cdot$ as suggested in [10; 4.8, p. 18] was used by Leicht [6].

Added in Proof (December 5,1968): Details are given in the "Anhang" of [2].

REFERENCES

[1] H.-B. Brinkmann, "Relations for Exact Categories", to appear (Preprint 1968).

[2] H.-B. Brinkmann und D. Puppe, "Exakte und Abelsche Kategorien, Korrespondenzen", Springer Lecture Notes in Mathematics, to appear.

[3] M.S. Calenko, "Relations for Quasiexact Categories", Mat. Sbornik (N.S.) 73(115); 564-584 (1967). In Russian.

[4] P. Hilton, "Correspondences and Exact Squares", Proc. Conf. Cat. Alg. 1965, (La Jolla, California); 254-271, Springer 1966.

[5] J. Lambek, "Goursat's Theorem and the Zassenhaus Lemma", Canadian J. Math., 10; 45-56 (1958).

[6] J.B. Leicht, "Remarks on the Axiomatic Theory of Additive Relations", Unpublished Manuscript (1964).

[7] J.L. MacDonald, "Coherence of Adjoints, Associativities and Identities", Arch. Math., 19; 398-401 (1968).

[8] S. MacLane, "Natural Associativity and Commutativity", Rice Univ. Studies, 49 (4); 28-46 (1963).

[9] B. Mitchell, "Theory of Categories", Academic Press 1965.

[10] D. Puppe, "Korrespondenzen in Abelschen Kategorien", Math. Ann., 148; 1-30 (1962).

GALOIS OBJECTS AND EXTENSIONS OF HOPF ALGEBRAS

By

Stephen U. Chase*

Let R be a commutative ring with unit, and A be a commutative, cocommutative Hopf R-algebra with antipode which is a finitely generated projective R-module (when discussing coalgebras and Hopf algebras we shall use the notation and terminology of [21]). Our principal result is a natural isomorphism -

$$X(A) \approx \mathrm{Ext}^1_{\underline{S}}(\underset{\sim}{A}^*, U) \tag{1}$$

where: (a) X(A) is the group of isomorphism classes of Galois A-objects, to be defined later, (b) $\underline{S}$ is the category of abelian sheaves relative to a suitably chosen Grothendieck topology on the category of commutative R-algebras, (c) $\underset{\sim}{A}^*$ is the sheaf represented by the linear dual A^* of A, and (d) U is the sheaf which assigns to each commutative R-algebra its multiplicative group of invertible elements. The coverings in our topology are essentially "Zariski" coverings; i.e., they arise from certain rings of fractions of R.

---

* The author wishes to acknowledge the support of N.S.F. GP-7945, Battelle Memorial Institute, Cornell University, and the Alfred E. Sloan Foundation. Detailed proofs of the results presented in this manuscript will appear in a subsequent issue of the Springer Lecture Notes, ETH Series.

The isomorphism (1) assumes a somewhat more classical form in the following special case. Let n be prime to the characteristic of a field k, $K = k(\zeta)$ with ζ a primitive n^{th} root of 1, U_n be the group of nth roots of 1 in K, and Π and Γ be the Galois groups of k^s/k and K/k, respectively, with k^s a separable closure of k. Then, for any finite abelian group J of exponent n -

$$\mathrm{Hom}_c(\Pi, J) \approx \mathrm{Ext}^1_{\mathbb{Z}\Gamma}(\mathrm{Hom}_{\mathbb{Z}}(J, U_n), U(K)) \tag{2}$$

where the left-hand side denotes continuous homomorphisms from the compact topological group Π to the discrete group J, $U(K)$ is the multiplicative group of invertible elements of K, and $\mathrm{Hom}_{\mathbb{Z}}(J,U_n)$ receives its Γ-module structure from that of U_n. If k contains a primitive n^{th} root of 1, then $K = k$ and $\Gamma = \{1\}$ and (2) becomes -

$$\mathrm{Hom}_c(\Pi, J) \approx \mathrm{Ext}^1_{\mathbb{Z}}(\mathrm{Hom}_{\mathbb{Z}}(J, U_n), U(k)) \tag{3}$$

which is, in essence, the well-known Kummer isomorphism of field theory (see, e.g., [1, pp. 19-22] or [19, Chapitre X, §3(b), p. 163]).

The isomorphism (1) may be viewed as an analogue, for finite commutative group schemes, of the Weil-Barsotti formula for abelian varieties ([18, Chapitre VII, Theorem 6, p. 184] and [14, Chapter III, 18]); it is also related to the Cartier-Shatz formula of [17, Proposition 1, p. 413]. It reformulates and generalizes some of the work of H. Hasse [12],

P. Wolf [23], D. K. Harrison [10], M. Orzech [15, 16], H. Epp [8] and others on Galois algebras and Kummer theory. The proof relies heavily on coalgebraic techniques, and bears some relation to the method used by Chase and Rosenberg in [6].

Galois objects form a ring-theoretic analogue of the principal homogeneous spaces of geometry, both notions being special cases of J. Giraud's torseurs in a topos [9]. We use a naive definition which is sufficient for our purposes. Let R be as above, and A be a commutative Hopf R-algebra with antipode. An <u>A-object</u> is a pair (S,α), where S is a commutative R-algebra and $\alpha: S \longrightarrow S \otimes A$ is an R-algebra homomorphism which gives to S the structure of a right A-comodule (unadorned $\otimes$ means $\otimes_R$). For brevity we shall denote the pair (S,α) by the symbol S alone, writing $\alpha = \alpha_S$ when the map α needs explicit mention. We define the algebra homomorphism $\gamma_S: S \otimes S \longrightarrow S \otimes A$ by the formula -

$$\gamma_S(x \otimes y) = (x \otimes 1)\alpha_S(y) \qquad (x,y \text{ in } S) \tag{4}$$

S will be called a <u>Galois A-object</u> if the following conditions hold -

$$S \text{ is a faithfully flat R-module.} \tag{5a}$$

$$\gamma_S: S \otimes S \longrightarrow S \otimes A \text{ is an isomorphism.} \tag{5b}$$

Before considering several examples of Galois objects, we introduce a notational device which is quite useful when dealing with coalgebras and comodules [21]. For x in a coalgebra C with comultiplication $\Delta_C: C \longrightarrow C \otimes C$, we write

$\Sigma_{(x)} x_{(1)} \otimes x_{(2)}$ to denote $\Delta_C(x)$, $\Sigma_{(x)} x_{(1)} \otimes x_{(2)} \otimes x_{(3)}$ to denote $(\Delta_C \otimes 1)\Delta_C(x) = (1 \otimes \Delta_C)\Delta_C(x)$, etc. If $f: C \otimes \ldots \otimes C \longrightarrow M$ is an R-module homomorphism, we write -

$$\sum_{(x)} f(x_{(1)}, \ldots, x_{(n)}) = f(\sum_{(x)} x_{(1)} \otimes \ldots \otimes x_{(n)})$$

In similar fashion, if A is a commutative Hopf algebra and S is an A-object, we write $\Sigma_{(x)} x_{(1)} \otimes x_{(2)}$ to denote $\alpha_S(x)$, $\Sigma_{(x)} x_{(1)} \otimes x_{(2)} \otimes x_{(3)}$ to denote $(\alpha_S \otimes 1)\alpha_S(x) = (1 \otimes \Delta_A)\alpha_S(x)$, and so on. The formula for $\gamma_S: S \otimes S \longrightarrow S \otimes A$ then becomes -

$$\gamma_S(x \otimes y) = \sum_{(y)} xy_{(1)} \otimes y_{(2)} \qquad (x, y \text{ in } S)$$

Suppose now that A is a Hopf R-algebra and a finitely generated projective R-module, in which case A will be called a <u>finite</u> Hopf algebra. Then $A^* = \mathrm{Hom}_R(A,R)$ is likewise a finite Hopf R-algebra [17, p. 413] and is cocommutative if and only if A is commutative. In this case, if S is an A-object, we have the natural isomorphisms -

$$\mathrm{Hom}_R(S, S \otimes A) \approx \mathrm{Hom}_R(S, \mathrm{Hom}_R(A^*, S)) \approx \mathrm{Hom}_R(A^* \otimes S, S) \qquad (6)$$

the first arising from our hypotheses on A and the second being the usual adjointness isomorphism. Then $\alpha_S: S \longrightarrow S \otimes A$, an element of the left-hand side, corresponds to a map $\beta_S: A^* \otimes S \longrightarrow S$ in the right-hand side. The fact that α_S gives to S a right A-comodule structure yields easily that β_S gives to S a left A^*-module structure. We shall write $\beta_S(u \otimes x) = u(x)$ for u in A^*, x in S. Routine computations establish the formulae below -

$$u(x) = \sum_{(x)} x_{(1)} \langle u, x_{(2)} \rangle \tag{7a}$$

$$u(xy) = \sum_{(u)} u_{(1)}(x) u_{(2)}(y) \tag{7b}$$

for x,y in S and u in A^*, where $\langle\rangle: A^* \otimes A \longrightarrow R$ is the duality pairing. (7b), which links the algebra structure of S to the coalgebra structure of A^*, is equivalent to the condition that α_S be a homomorphism of R-algebras; it implies that A^* measures S to S in the sense of [21, p. 265].

We turn now to some examples of Galois A-objects. If G is a group, we shall denote by RG the group algebra of G with coefficients in R, a cocommutative Hopf R-algebra with the usual comultiplication and counit (i.e., augmentation). The antipode $\lambda_{RG}: RG \longrightarrow RG$ is defined by $\lambda_{RG}(\sigma) = \sigma^{-1}$ for σ in G. If G is finite, in which case RG is a finite Hopf R-algebra, we shall write $GR = (RG)^*$; note that GR is simply the set of functions from G to R with the pointwise algebra operations, and coalgebra operations and antipode arising in an easily-ascertained fashion from the group operations in G.

If S is a GR-object, with G finite, then S may be viewed, via (7a), as a left module over $RG = (GR)^*$. (7b) then guarantees that the elements of G act as R-algebra automorphisms of S. If, conversely, S is a commutative R-algebra on which G acts via R-algebra automorphisms, then S is in the usual way a left RG-module, and the map $\alpha_S: S \longrightarrow \otimes GR$ corresponding, via (6), to the structure map $RG \otimes S \longrightarrow S$

renders S a GR-object. Finally, scrutiny of [4, Theorem 1.3e] and [6, Lemma 2.5] renders apparent the fact that S is a Galois GR-object if and only if S is a Galois extension with Galois group G in the sense of [4]. In particular, if R and S are fields, S is a Galois GR-object if and only if S is a normal, separable extension of R with Galois group G ; this assertion admits an easy direct proof.

Having described the Galois GR-objects, at least for G finite, one is tempted to inquire about the RG-objects. We consider the special case in which $G = \mathbb{Z}$, the infinite cyclic group. If we write $\mathbb{Z}$ multiplicatively with generator t, then $R\mathbb{Z}$ is a free R-module with basis consisting of all integral powers of t. If S is an $R\mathbb{Z}$-object, we define S_n for any integer n by the condition -

$$S_n = \{x \text{ in } S \mid \alpha_S(x) = x \otimes t^n \text{ in } S \otimes R\mathbb{Z}\}.$$

It is then easily verified that $S = \Sigma_{-\infty}^{+\infty} \oplus S_n$ and $S_m S_n \subseteq S_{m+n}$; i.e., S is a $\mathbb{Z}$-graded R-algebra with S_n as its nth homogeneous component. If, conversely, $S = \Sigma_{-\infty}^{+\infty} \oplus S_n$ is $\mathbb{Z}$-graded R-algebra, then S becomes an $R\mathbb{Z}$-object with $\alpha_S\colon S \longrightarrow S \otimes R\mathbb{Z}$ defined by the formula -

$$\alpha_S(x) = x \otimes t^n \qquad (x \text{ in } S_n)$$

Thus the $R\mathbb{Z}$-objects are precisely the $\mathbb{Z}$-graded R-algebras. Finally, it can be shown that, if S is a Galois $R\mathbb{Z}$-object, then S_1 is a projective R-module of rank one [3, Chapitre 2, §4];

in fact, the correspondence $S \longrightarrow S_1$ is one-to-one between the Galois $R\mathbb{Z}$-objects and the projective R-modules of rank one. This correspondence is discussed, for example, in [20, Exposé 8, §2.3.1].

For our last example, let us replace R by a field k of characteristic $p > 0$, and consider the k-algebra $A = k[t]/(t^{p^n})$, n a positive integer. Let z be the image of t in A; then z generates A as a k-algebra and $z^{p^n} = 0$. One sees immediately that A can be given, in a unique way, the structure of a finite commutative Hopf k-algebra such that $\Delta_A(z) = z \otimes 1 + 1 \otimes z$, $\varepsilon_A(z) = 0$, and $\lambda_A(z) = -z$. Now let $S = k(x)$ be a purely inseparable field extension of k, with x^{p^n} in k, $x^{p^{n-1}}$ not in k. It is then easily verified that S is a Galois A-object if $\alpha_S: S \longrightarrow S \otimes A$ is defined by the formula -

$$\alpha_S(x) = x \otimes 1 + 1 \otimes z$$

More complicated examples of this type have been considered by M. E. Sweedler, who has proved a theorem which implies the following fact: If K is any finite purely inseparable field extension of k, then there is a field S containing K which is a Galois A-object for some finite commutative Hopf k-algebra A [22, Theorem 6].

Next we describe the group X(A) introduced in (1), for A a commutative, cocommutative Hopf R-algebra. If S and S′ are A-objects, an R-algebra isomorphism $f: S \xrightarrow{\approx} S'$

will be called an A-isomorphism if $\alpha_{S'} f = (f \otimes 1)\alpha_S : S \longrightarrow S' \otimes A$. The elements of $X(A)$ are the A-isomorphism classes of Galois A-objects. If S_1, S_2 are Galois A-objects, we set -

$$cl(S_1) + cl(S_2) = cl(S)$$

(cl( ) meaning "A-isomorphism class of ( )"), where S is obtained as follows. Define maps $\theta, \psi : S_1 \otimes S_2 \longrightarrow S_1 \otimes S_2 \otimes A$ by the formulae -

$$\theta(x \otimes y) = \sum_{(y)} x \otimes y_{(1)} \otimes y_{(2)}$$

$$\psi(x \otimes y) = \sum_{(x)} x_{(1)} \otimes y \otimes x_{(2)}$$

for x in S_1, y in S_2, and let $S = \{w \text{ in } S_1 \otimes S_2 / \theta(w) = \psi(w)\}$. The coassociativity of $\alpha_{S_1}, \alpha_{S_2}$ guarantees that the common restriction to S of θ and ψ factors uniquely through $S \otimes A$, thus producing a map $\alpha_S : S \longrightarrow S \otimes A$. Using the cocommutativity of A, it can then be shown that $S = (S, \alpha_S)$ is a Galois A-object; furthermore, the operation "+" thereby obtained is well-defined and renders $X(A)$ an abelian group. The zero element of $X(A)$ is $cl(A)$, where $\alpha_A = \Delta_A : A \longrightarrow A \otimes A$.

Suppose now that $\varphi : B \longrightarrow A$ is a homomorphism of commutative, cocommutative Hopf R-algebras. If S is a Galois A-object, we define R-algebra homomorphisms $\zeta, \omega : S \otimes B \longrightarrow S \otimes A \otimes B$ by the formulae -

$$\zeta(x \otimes b) = \sum_{(b)} x \otimes \varphi(b_{(1)}) \otimes b_{(2)}$$

$$\omega(x \otimes b) = \sum_{(x)} x_{(1)} \otimes x_{(2)} \otimes b$$

and set $\tilde{\varphi}(S) = \{z \text{ in } S \otimes B | \zeta(z) = \omega(z)\}$. $\tilde{\varphi}(S)$ is an R-subalgebra of $S \otimes B$, and it is easily verified that the restriction to $\tilde{\varphi}(S)$ of the map $S \otimes \Delta_B\colon S \otimes B \longrightarrow (S \otimes B) \otimes B$ factors uniquely through $\tilde{\varphi}(S) \otimes B$, thus producing a map $\alpha_{\tilde{\varphi}(S)}\colon \tilde{\varphi}(S) \longrightarrow \tilde{\varphi}(S) \otimes B$. It can then be shown that $\tilde{\varphi}(S) = (\tilde{\varphi}(S), \alpha_{\tilde{\varphi}(S)})$ is a Galois B-object; moreover, the map $X(\varphi)\colon X(A) \longrightarrow X(B)$ defined by $X(\varphi)(cl(S)) = cl(\tilde{\varphi}(S))$ is a homomorphism of abelian groups. Routine computations then establish that X is an additive contravariant functor from the category $\underline{H}$ of commutative, cocommutative Hopf R-algebras (with antipode) to the category of abelian groups.

The group $X(A)$ was treated, from a similar point of view, by D. K. Harrison in [11], for the special case in which $A = JR$, J a finite abelian group.

We describe $X(A)$ below for some of the special cases previously considered.

Let k be a field, and J be a finite abelian group. Then -

$$X(Jk) \approx \mathrm{Hom}_c(\Pi, J) \tag{8a}$$

where Π is the Galois group of a separable closure k^s of k, and the right-hand side denotes continuous homomorphisms from the compact group Π to the discrete group J. The isomorphism is natural in J, and may be described explicitly as follows. Let $\chi\colon \Pi \longrightarrow J$ be a continuous homomorphism, and S_χ be the k-algebra of all functions $u\colon J \longrightarrow k^s$ such that $\pi(u(\sigma)) = u(\chi(\pi)\sigma)$ for all π in Π, σ in J. The formula -

$$\tau(u)(\sigma) = u(\sigma\tau) \qquad (u \text{ in } S_\chi;\ \sigma,\tau \text{ in } J)$$

defines an action of J on S_χ via k-algebra automorphisms, in virtue of which S_χ becomes a Jk-object, as explained earlier. A routine exercise in the Galois theory of fields establishes that S_χ is a Galois Jk-object. The isomorphism introduced above then sends χ to $cl(S_\chi)$ in X(Jk).

If R is a commutative ring, then the previously discussed correspondence between Galois $R\mathbb{Z}$-objects and projective R-modules of rank one establishes an isomorphism -

$$X(R\mathbb{Z}) \approx Pic(R) \tag{8b}$$

the latter denoting the Picard group of isomorphism classes of projective R-modules of rank one [3, Chapitre 2, §4].

If R is a commutative ring and $\mathbb{Z}_n = \mathbb{Z}|n\mathbb{Z}$, then there exists an exact sequence -

$$U(R) \xrightarrow{(\ )^n} U(R) \longrightarrow X(R\mathbb{Z}_n) \longrightarrow Pic(R) \xrightarrow{(\ )^n} Pic(R) \tag{8c}$$

where U(R) is the multiplicative group of invertible elements of R, and the left-most and right-most maps send an element to its nth power.

These facts are well-known in the context of principal homogeneous spaces. A somewhat more detailed treatment of (8a), in a more general situation, can be found in [7, §3].

(8b) is discussed, for example, in [20, Exposé 8, §2.3.1]. (8c) can be established by an easy direct argument; it can also be obtained from (8b) and standard cohomological techniques.

We shall motivate our approach to the isomorphism (1) by a brief consideration of the special case provided by the isomorphism (3). Let k be a field of characteristic prime to a given natural number n, and assume that k contains all n^{th} roots of 1. If J is a finite abelian group of exponent n, the isomorphism (3) determines, and is determined by, an isomorphism -

$$X(Jk) \approx \mathrm{Ext}^1_{\mathbb{Z}}(\mathrm{Hom}(J,U_n),\ U(k)) \tag{9}$$

In order to define this map, let S be a Galois Jk-object, hence a commutative k-algebra on which J acts via k-algebra automorphisms. Let $V(S)$ be the set of all x in $U(S)$ such that $x^{-1}\sigma(x)$ is in k for all σ in J. $V(S)$ is clearly a subgroup of $U(S)$ which contains $U(k)$. If x is in $V(S)$, we define a function φ_x on J by the formula -

$$\varphi_x(\sigma) = x^{-1}\sigma(x) \qquad (\sigma \text{ in } J)$$

φ_x takes values in $U(k)$; furthermore, we see that, for σ,τ in J -

$$\begin{aligned}\varphi_x(\sigma\tau) &= x^{-1}\sigma\tau(x) = x^{-1}\sigma\{x(x^{-1}\tau(x))\} = (x^{-1}\sigma(x))(x^{-1}\tau(x) \\ &= \varphi_x(\sigma)\varphi_x(\tau)\end{aligned}$$

whence φ_x is a homomorphism. Since J has exponent n, it follows that φ_x takes values in the subgroup U_n of nth roots of 1 in k. Also, the fact that S is a Galois JK-object guarantees that the subring of S left fixed by all elements of J is precisely k, whence $\varphi_x(\sigma) = 1$ for all σ in J if and only if x is in $U(k)$.

Suppose now that $\varphi: J \longrightarrow U_n$ is an arbitrary homomorphism. It is then clear that, with regard to the Galois cohomology of, for example, [19, Chapitre X], φ is a one-cocycle of J with values in $U(S)$, hence a coboundary by a slight generalization of Hilbert's Theorem 90 [19, Proposition 2, p. 158]; i.e., there exists x in $U(S)$ such that $\varphi(\sigma) = x^{-1}\sigma(x)$ for all σ in J. x is then in $V(S)$, and $\varphi = \varphi_x$. We may then conclude that the sequence below is exact -

$$\xi_S: 0 \longrightarrow U(k) \longrightarrow V(S) \longrightarrow \mathrm{Hom}_{\mathbb{Z}}(J,U_n) \longrightarrow 0$$

(where the arrows denote the inclusion map and the map $x \longrightarrow \varphi_x$, respectively), and therefore defines an element of $\mathrm{Ext}^1_{\mathbb{Z}}\{\mathrm{Hom}_{\mathbb{Z}}(J,U_n), U(k)\}$.

Of course, the extension ξ_S determines, and is determined by, the pairing $V(S) \times J \longrightarrow U_n$ defined by $(x,\sigma) \longrightarrow \varphi_x(\sigma) = x^{-1}\sigma(x)$. Now, since J has exponent n and k contains all n^{th} roots of 1 in S, it is easy to see that $V(S)$ is precisely the set of non-zero elements of S whose n^{th} powers are in k. Thus, for the special case in which S is a normal, separable field extension of k with Galois group J, the pairing introduced above coincides with that arising in the classical formulation of Kummer theory (see, e.g., [1, pp. 19-22]). The mapping (9) assigns to $cl(S)$ in $X(Jk)$ the element $cl(\xi_S)$ in $Ext^1_{\mathbb{Z}}(Hom_{\mathbb{Z}}(J,U_n), U(k))$, and the techniques of [1, pp. 19-22] can then be used to show that this mapping is a well-defined isomorphism of abelian groups. The relation between (9) and (3) becomes more apparent if one observes that $Hom_{\mathbb{Z}}(J,U_n) \approx \underline{A}(kJ,k) = \underline{A}((Jk)^*,k)$, $\underline{A}(\ )$ denoting k-algebra homomorphisms.

In order to obtain the isomorphism (1), we proceed in a quite similar fashion. Let A be a commutative, cocommutative finite Hopf R-algebra with antipode, and S be a Galois A-object. We define $V(S)$ to be the set of all x in $U(S)$ such that $x^{-1}u(x)$ is in R for all u in A^*, with $u(x)$ as in (7a). $V(S)$ is a subgroup of $U(S)$ containing $U(R)$. If x is in $V(S)$, we define a mapping $\varphi_x: A^* \longrightarrow R$ by the formula -

$$\varphi_x(u) = x^{-1}u(x) \qquad (u \text{ in } A^*)$$

Routine computations, entirely similar to those of the special case just discussed, show that φ_x is a homomorphism of R-algebras, and the sequence below is exact -

$$0 \longrightarrow U(R) \longrightarrow V(S) \longrightarrow \underline{A}(A^*,R) \qquad (10)$$

where $\underline{A}(\ )$ denotes R-algebra homomorphisms, and the unlabeled arrows denote the inclusion map and the map $x \longrightarrow \varphi_x$, respectively. In this case the latter map is not in general surjective.

Now, the left A^*-module structure on S induces, in the usual way, a right A^*-module structure on $S^* = \mathrm{Hom}_R(S,R)$. For the special case in which $S^* \approx A^*$ as right A^*-modules, the surjectivity of the map $V(S) \longrightarrow \underline{A}(A^*,R)$ may be seen as follows. If z generates S^* as an A^*-module and $\varphi\colon A^* \longrightarrow R$ is an R-algebra homomorphism, we define $f\colon S^* \longrightarrow R$ by the equation

$$f(zu) = \varphi(u) \qquad (u \text{ in } A^*) \ .$$

S is a finitely generated projective R-module, and thus there exists a unique x in S such that $f = \langle \text{---}, x\rangle$, where $\langle\ \rangle\colon S^* \otimes S \longrightarrow R$ is the duality pairing. It can then be shown that x is in $V(S)$ and $\varphi = \varphi_x$.

Although in general S^* need not be isomorphic to A^* as a right A^*-module, it is at least a projective A^*-module of rank one. The techniques of, for example, [5, §5] can then be used to show that $S^* \otimes T \approx A^* \otimes T$ as right $A^* \otimes T$-modules for some faithfully flat commutative R-algebra T. In fact, T can be chosen to be of the form -

$$T = \prod_{i=1}^{r} R_{x_i} \tag{11}$$

where $x_1,\ldots,x_r$ are elements of R, none of which are in the Jacobson radical, such that $x_1 + \ldots + x_r = 1$, and R_i is the localization of R at the multiplicatively closed subset generated by x_i. The preceding paragraphs then guarantee easily that a base-change from R to T renders the sequence (10) a short exact sequence of abelian groups.

It is thus natural to consider short exact sequences of sheaves in the following context. Given a commutative R-algebra S, we define a covering of S to be an R-algebra homomorphism of the form $S \longrightarrow S \otimes T$, where T is of the form (11) and x in S goes to $x \otimes 1$ in $S \otimes T$. One checks easily that this collection of maps satisfies axioms dual to those of a Grothendieck topology (for the definition and elementary properties of which we refer the reader to [2]), and thus gives rise to a Grothendieck topology on the category dual to the category $\underline{A}$ of commutative R-algebras. We may then consider the category $\underline{S}$ of abelian sheaves in this topology. An object of $\underline{S}$ is a (covariant) functor F from $\underline{A}$ to abelian groups such that, for any covering $d: S \longrightarrow S \otimes T$, the sequence below is exact -

$$0 \longrightarrow F(S) \xrightarrow{F(d)} F(S \otimes T) \xrightarrow{F(d^0) - F(d^0)} F(S \otimes T \otimes T)$$

with $d^i: S \otimes T \longrightarrow S \otimes T \otimes T$ $(i = 0,1)$ defined by $d^0(s \otimes t) = s \otimes 1 \otimes t$, $d^1(s \otimes t) = s \otimes t \otimes 1$. A map in $\underline{S}$ is simply a natural transformation of functors.

$\underline{S}$ is an abelian category, and a sequence -

$$0 \longrightarrow F' \xrightarrow{\psi} F \xrightarrow{\rho} F'' \longrightarrow 0$$

in $\underline{S}$ is a short exact sequence if and only if the following conditions hold -

The sequence $0 \longrightarrow F'(S) \xrightarrow{\Psi(S)} F(S) \xrightarrow{\rho(S)} F''(S)$ (12a) is an exact sequence of abelian groups for any object S of $\underline{A}$.

Given any object S of $\underline{A}$ and x'' in $F''(S)$, (12b)
there is a covering $d: S \longrightarrow S \otimes T$ and an x in $F(S \otimes T)$ such that $\rho(S \otimes T)(x) = F''(d)(x'')$.

For the proofs of these facts we refer to [2, Chapter II, §1]. Note finally that, since $\underline{S}$ is an abelian category, we may define $\operatorname{Ext}^{*}_{\underline{S}}(-,-)$ as, for example, in [13, Chapter XII]. In particular, if F and G are abelian sheaves, then $\operatorname{Ext}^{1}_{\underline{S}}(F,G)$ may be viewed as the abelian group of equivalence classes of short exact sequences in $\underline{S}$ of the form -

$$0 \longrightarrow G \longrightarrow E \longrightarrow F \longrightarrow 0$$

Now, given A and S as in (10) and T an object of $\underline{A}$, we note that $A \otimes T$ is a finite commutative, cocommutative Hopf T-algebra with antipode, and $S \otimes T$ is a Galois $A \otimes T$-object. It then follows easily from the preceding discussion that the sequence -

$$\xi_S\colon 0 \longrightarrow U \longrightarrow V_S \longrightarrow \underline{A}^{*} \longrightarrow 0$$

is a short exact sequence in $\underline{S}$; where, for T an object of $\underline{A}$, the abelian group $V_S(T) = V(S \otimes T)$ and the maps
$U(T) \longrightarrow V(S \otimes T)$ and
$V(S \otimes T) \longrightarrow \underline{A}^{*}(T) = \underline{A}(A^{*},T) \approx \underline{A}_T(\operatorname{Hom}_T(A \otimes T,\ T),T)$ ($\underline{A}_T$ denoting T-algebra homomorphisms) are defined as in (10), except that the roles of R, A, S are now being played by T, $A \otimes T$, and $S \otimes T$, respectively. The mapping (1) sends $cl(S)$ in $X(A)$

to $cl(\xi_S)$ in $\mathrm{Ext}^1_{\underline{S}}(\underset{\sim}{A}^*,U)$, and computations of an essentially routine nature ensure that it is a homomorphism of abelian groups.

In order to obtain the inverse map $\mathrm{Ext}^1_{\underline{S}}(\underset{\sim}{A}^*,U) \longrightarrow X(A)$, we turn to considerations of a more Hopf algebraic nature. Given a short exact sequence -

$$\xi\colon 0 \longrightarrow U \longrightarrow E \longrightarrow \underset{\sim}{A}^* \longrightarrow 0$$

in $\underline{S}$, we observe that the functor U is represented by the Hopf algebra $R\mathbb{Z}$. The techniques of faithfully flat descent then guarantee that E is likewise representable by some Hopf R-algebra H (see, for example, [14, p. III. 17-6]), whence ξ arises from a sequence -

$$A^* \xrightarrow{\rho} H \xrightarrow{\Psi} R\mathbb{Z}$$

of abelian cogroup objects in $\underline{A}$; i.e., commutative, cocommutative Hopf R-algebras with antipode. It is then easy to see that H is an $R\mathbb{Z}$-object, with $\alpha_H\colon H \longrightarrow H \otimes R\mathbb{Z}$ the composite map -

$$H \xrightarrow{\Delta_H} H \otimes H \xrightarrow{H \otimes \Psi} H \otimes R\mathbb{Z}$$

Therefore, as remarked earlier, H is a $\mathbb{Z}$-graded R-algebra -

$$H = \sum_{-\infty}^{+\infty} \otimes H_n$$

with $H_n = \{z \text{ in } H/\alpha_H(z) = z \otimes t^n\}$. Since α_H is a Hopf algebra homomorphism, it follows that each H_n is a sub-coalgebra of H, whence H_0 is even a Hopf subalgebra of H. Moreover, the image of ρ lies in H_0, thus providing a Hopf algebra homomorphism $\rho: A^* \longrightarrow H_0$. We then obtain a coalgebra homomorphism -

$$H_1 \otimes A^* \xrightarrow{H_1 \otimes \rho} H_1 \otimes H_0 \longrightarrow H_1 \tag{13}$$

the unlabeled arrow denoting the appropriate restriction of the multiplication map of H. Finally, H_1 is a finitely generated projective R-module, whence dualization of (13) yields an R-algebra homomorphism $\alpha_S: S \longrightarrow S \otimes A$, with $S = H_1^*$. It turns out that $S = (S, \alpha_S)$ is a Galois A-object, and the map (1) then sends $cl(\xi)$ in $Ext^1_{\underline{S}}(\underline{A}^*, U)$ to $cl(S)$ in $X(A)$. This completes our sketch of the construction of the isomorphism (1).

A final remark regarding the Grothendieck topology which we use to define our sheaves. Note that the coverings in this topology are of a very restricted type; in particular, the topology is much coarser than, for example, the faithfully flat topology on $\underline{A}$. Indeed, for the special case in which R is a local ring, it is trivially verified that every presheaf on $\underline{A}$ (i.e., covariant functor from $\underline{A}$ to abelian groups) is a sheaf. This is probably the most important step in the deduction of the isomorphism (2) from the isomorphism (1).

The foregoing material is a resume of a lecture presented at the Battelle conference on categorical and homological algebra, Seattle, June-July 1968. During the conference,

S. Shatz and D. Quillen suggested an alternate approach to the isomorphism (1), the ingredients of which include the Cartier-Shatz formula, a spectral sequence argument, and a well-known cohomological classification of principal homogeneous spaces. This method is similar to that of H. Epp [8], who derived (1), using sheaves in the faithfully flat topology, for the special case of group schemes whose duals are of multiplicative type (i.e., A has the property that $A \otimes T \approx JT$ for some faithfully flat commutative R-algebra T and finitely generated abelian group J).

REF

[1] Artin, E., and Tate, J., *Class Field Theory*, Princeton University mimeographed notes, Princeton, New Jersey (1960).

[2] Artin, M., *Grothendieck Topologies*, Harvard University mimeographed notes, Cambridge, Massachusetts (1962).

[3] Bourbaki, N., *Algèbre Commutative, Chapters I-II*, Hermann, Paris (1962).

[4] Chase, S. U., Harrison, D. K., and Rosenberg, Alex, "Galois Theory and Galois Cohomology of Commutative Rings," *Memoirs Amer. Math. Soc.*, Vol. 52; 15-33, (1965).

[5] Chase, S. U., and Rosenberg, Alex, "Amitsur Cohomology and the Brauer Group," *Memoirs Amer. Math. Soc.*, Vol. 52; 34-79, (1965).

[6] Chase, S. U., and Rosenberg, Alex, "A Theorem of Harrison, Kummer Theory, and Galois Algebras," *Nagoya Math. J.*, Vol. 27; 663-685, (1966).

[7] Chase, S. U., "Abelian Extensions and a Cohomology Theory of Harrison," 375-403 in *Proceedings of the Conference on Categorical Algebra, La Jolla (1965)*, Springer-Verlag, New York Inc., (1966).

[8] Epp, H. P., *Commutative Group Schemes, Harrison's Theorem, and Galois Extensions*, Ph.D. thesis, Northwestern University, (1966).

[9] Giraud, J., *Cohomologie Non-Abelienne*, Columbia University Notes (1965).

[10] Harrison, D. K., "Abelian Extensions of Arbitrary Fields," *Trans. Amer. Math. Soc.*, Vol. 106; 230-235, (1963).

[11] __________, "Abelian Extensions of Commutative Rings," *Memoirs Amer. Math. Soc.* 52; 1-14, (1965).

[12] Hasse, H., "Invariante Kennzeichnung Galoisscher Körper mit Vorgegebener Galoisgruppe," *J. Reine Angew. Math.*, Vol. 187; 14-43, (1950).

[13] MacClane, S., *Homology*, Academic Press, New York, (1963).

[14] Oort, F., *Commutative Group Schemes*, (Lecture Notes in Mathematics, Vol. 15), Springer-Verlag, Berlin, (1966).

[15] Orzech, M., "A Cohomology Theory for Commutative Galois Extensions," *Math. Zeitsch.*, Vol. 105; 128-140, (1968).

[16] __________, "A Cohomological Description of Abelian Galois Extensions," to appear.

[17] Shatz, S., "Cohomology of Artinian Group Schemes Over Local Fields," *Ann. of Math.*, Vol. 79; 411-449, (1964).

[18] Serre, J. P., *Groupes Algebriques et Corps de Classes*, Hermann, Paris, (1959), (Act. Sci. Ind. #1264).

[19] __________, *Corps Locaux*, Hermann, Paris, (1962), (Act. Sci. Ind. #1296).

[20] Strasbourg University Department of Mathematics, "Groupes Algebriques," *Seminaire Heidelberg-Strasbourg Annee 1965-66*.

[21] Sweedler, M. E., "The Hopf Algebra of an Algebra as Applied to Field Theory," *J. of Algebra*, Vol. 8; 262-276, (1968).

[22] __________, "Structure of Inseparable Extensions," *Annals of Math.*, Vol. 87; 401-410, (1968).

[23] Wolf, P., "Algebraische Theorie der Galoisschen Algebren," *Deutscher Verlag der Wissenschaften, Math. Forschungsberichte III*, Berlin, (1956).

THREE DIMENSIONAL NON-ABELIAN COHOMOLOGY FOR GROUPS *

by

Paul Dedecker

So far non-abelian cohomology has been discussed in dimensions $n \leq 2$. The following is an effort to discuss dimension $n = 3$ and solves the main difficulties so that just polishing work remains necessary. However an n-dimensional theory for $n > 3$ remains a remote target.

The present report is a direct continuation of [2].

I am grateful for the hospitality of the Center for Advanced Studies, Wesleyan University, during the final phase of the redaction. Typing was kindly offered by the University of Kentucky.

1. Introduction.

We shall assume the reader is familiar with the n-dimensional theory, $n \leq 2$, as developed in earlier papers since 1958 [1], [2], [3]. Let us however recall a few facts. Let G be an arbitrary group. Then the category $\mathcal{M}_1$ of G-groups is satisfactory to serve as coefficient category for a cohomology theory of G only in dimensions 0 and 1. It is not so in dimension 2 and that is the reason why the theory took so much time to take off. As I have shown since 1958 we have to

* Research supported by NATO Research grant No. 224 and Centre belge d'Algèbre et de Topologie

replace $\mathcal{M}_1$ by the category $\mathcal{M}_2$ of crossed groups which has a forgetful functor $\mathcal{M}_2 \to \mathcal{M}_1$. This concept was introduced by J. H. C. Whitehead under the name of crossed module. Very roughly speaking a crossed group (or crossed module) is a group endowed with an additional structure related to its group of automorphisms. In dimension 3 we shall have to introduce a new category $\mathcal{M}_3$ whose objects we shall call supercrossed groups and which has a forgetful functor $\mathcal{M}_3 \to \mathcal{M}_2$. As is well known, if the functor $H^0(G, -)$ has values in groups, this is not the case of the functor $H^1(G, -)$ whose values are often described as pointed sets (e.g. [6] Serre). However this can be made much more precise since this functor has values in the category of polypi [1, I, II], [3]. Similarly the functor $H^2(G, -)$ takes its values in a category known as the category of spiders. So far the structure of values of the functor $H^3(G, -)$ is not sufficiently known to give these objects a name. I hope however to say enough to show that they are very curious and interesting animals with nice feathers.

Of course the usefulness of this cohomology theory is that (i) it is functorial and (ii) it produces a cohomology exact sequence associated with a short exact sequence in the coefficients category, a notion respected by the forgetful functors $\mathcal{M}_3 \to \mathcal{M}_2 \to \mathcal{M}_1$. Let us mention here that Giraud [5], [6] does not produce a truly functorial cohomology, nor a true exact cohomology-sequence. He has in fact a section

$$\cdots \to H^1(A'') \to H^2(A') \multimap H^2(A) \to H^2(A'')$$

associated to a short exact sequence of sheaves $A' \to A \to A''$. In there the symbol $—\!\circ$ *does not represent a map but a relation*. This trouble (showing that this kind of H^2 *is not functorial*) occurs when one does not go into a good category $\mathcal{M}_2$ of coefficients but wants to remain in $\mathcal{M}_1$. However this defect in Giraud's theory could be easily avoided and certainly a nice paper could be written clarifying this and making Giraud's beautiful theory more accessible. It would then be important to generalize our H^3 to the more general situation considered by Giraud.

2. *Quick review of 2-cohomology.*

We remind that a *crossed-group* A is a system $A = (H, \rho, \Pi, \Phi)$ in which H and Π are (arbitrary) groups, ρ is a homomorphism $\rho : H \to \Pi$ and Φ is an action of Π onto H by automorphisms, subject to conditions (i) and (ii) below. In these, if $\alpha \in \Pi$, $h \in H$, the result of α acting on (or twisting) h is denoted by ${}^{\alpha}h \in H$.

(i) If $h, k \in H$, then ${}^{\rho k}h = khk^{-1}$.

(ii) If $\alpha \in \Pi$, $h \in H$, then $\rho({}^{\alpha}h) = \alpha . \rho h . \alpha^{-1}$.

A morphism $A \to A'$ between two crossed groups is of course a pair of morphisms $j : H \to H'$, $\gamma : \Pi \to \Pi'$ compatible with the structures, i.e. making commutative the squares

$$(2.1) \qquad \begin{array}{ccc} \Pi & \xrightarrow{\gamma} & \Pi' \\ \uparrow{\scriptstyle\rho} & & \uparrow{\scriptstyle\rho'} \\ H & \xrightarrow[j]{} & H' \end{array} \qquad\qquad \begin{array}{ccc} \Pi \times H & \xrightarrow{\Phi} & H \\ {\scriptstyle\gamma}\downarrow \;\; \downarrow{\scriptstyle j} & & \downarrow{\scriptstyle j} \\ \Pi' \times H' & \xrightarrow[\Phi']{} & H' \end{array}$$

This defines the category $\mathcal{M}_2$ and the forgetful functor $\mathcal{M}_2 \to \mathcal{M}_1$. ($\mathcal{M}_1$ = the category of groups) is defined by

$$(H, \rho, \Pi, \Phi) \rightsquigarrow H \ .$$

Theorem 2.1. This forgetful functor has no section.

Roughly speaking, this means that it is not possible to associate to a group a group of automorphisms in a functorial way.

If the theorem were false, there would be no need of crossed groups to define non-abelian cohomology. To the best of my knowledge no proof of the theorem exists. However this non-proved theorem is responsible for the snaggy $\multimap$ in Giraud's theory.

For G an arbitrary group, we denote

$$\begin{aligned}
C^0(G, A) &= C^0(G, H) = H \ , && \text{0-cochains,} \\
C^1(G, A) &= C^1(G, H) = \underline{\mathrm{App}}(G, H) \ , && \text{1-cochains,} \\
C^2(G, A) &= \underline{\mathrm{App}}(G, \Pi) \times \underline{\mathrm{App}}(G \times G, H) \ , && \text{2-cochains,}
\end{aligned}$$

where App denotes the set of all maps in between the underlying sets. Moreover we denote by $Z^2(G, A)$ the set of 2-cochains (ϕ, h) such that $(s, t, u, \dots \in G)$:

(2.2.a) $$^{\phi(s)}h(t, u) \cdot h(s, tu) = h(s, t) \cdot h(st, u) \ ,$$

(2.2.b) $$\phi(s) \cdot \phi(t) = \rho h(s, t) \cdot \phi(st) \ .$$

An action

(2.3) $$* : C^1(G, A) \times Z^2(G, A) \to Z^2(G, A)$$

is defined by

$$a*(\phi, h) = (\phi', h'), \qquad a : G \to H, \quad (\phi, h) \in Z^2(G, A),$$

$$\phi'(s) = \rho a(s) \cdot \phi(s), \quad h'(s,t) = a(s) \cdot {}^{\phi(s)}a(t) \cdot h(s,t) \cdot a(st)^{-1}.$$

This is a group action, considering $C^1(G, A)$ as a group in the obvious way. The orbits form a set (thick 2-cohomology)

$$\mathbb{H}^2(G, A) = Z^2(G, A)/C^1(G, A).$$

There are other actions

(2.3) $$\square : \Pi \times C^1(G, A) \to C^1(G, A)$$

(2.4) $$\nabla : \Pi \times Z^2(G, A) \to Z^2(G, A)$$

which are defined by the following formulas in which

$$\alpha \in \Pi, \quad a \in C^1(G, A), \quad (\phi, h) \in Z^2(G, A):$$

$$\alpha \,\square\, a = a', \qquad a'(s) = {}^{\alpha}a(s);$$

$$\alpha \,\nabla (\phi, h) = (\bar{\phi}, \bar{h})$$

$$\bar{\phi}(s) = \alpha \cdot \phi(s) \cdot \alpha^{-1} \overset{\text{def}}{=} {}^{\alpha}\phi(s), \quad \bar{h}(s,t) = {}^{\alpha}h(s,t).$$

This satisfies the following "distributivity law"

$$\alpha \nabla [a * (\phi, h)] = (\alpha \square a) * (\alpha \nabla (\phi, h))]$$

which shows that the action ∇ transforms the whole orbit of (ϕ, h) under C^1 into the corresponding orbit of $\alpha \nabla (\phi, h)$. This thus defines an action of Π onto $\mathbb{H}^2(G, A)$. The set of orbits in $\mathbb{H}^2(G, A)$ under this last action is called the thin 2-cohomology and denoted

$$H^2(G, A) = \mathbb{H}^2(G, A)/\Pi .$$

We can also consider the crossed product (or semi-direct product)

$$\Gamma = C^1(G, A) \boxtimes \Pi$$

which has as underlying set the product $C^1(G, A) \times \Pi$ with multiplication law

$$(b, \beta) \cdot (a, \alpha) = (b \cdot (\beta \sqcap a), \beta\alpha) .$$

We then define a group action

$$\otimes : \Gamma \times Z^2(G, A) \to Z^2(G, A)$$

by

$$(a, \alpha) \otimes (\phi, h) = a * [\alpha \nabla (\phi, h)] .$$

This obviously produces an isomorphism

$$H^2(G, A) \cong Z^2(G, A)/\Gamma .$$

A 2-cocycle (ϕ, h) is said to be <u>neutral</u> if $h : G \times G \to H$ is the trivial map, which we shall denote by 1, taking every pair (s, t) into the unit element of H. In this case $\phi : G \to \Pi$ is a good standing homomorphism. A 2-dimensional cohomology class in $\mathbb{H}^2(G, A)$ or in $H^2(G, A)$ is said to be <u>neutral</u> if it contains a neutral 2-cocycle.

Let $\theta : G \to \Pi$ be a homomorphism. We then denote by

$$\mathbb{H}^1_\theta(G, A) = \underline{\mathrm{Hom}_\theta}(G, A) = Z^1_\theta(G, A) \subset C^1(G, A)$$

the set of θ-<u>crossed homomorphisms</u> $f : G \to H$, i.e. the set of maps $f : G \to A$

such that

$$f(st) = f(s) . {}^{\theta(s)}f(t) .$$

An action is defined:

$$* : H \times Z^1_\theta(G, A) \to Z^1_\theta(G, A) ,$$

by putting

$$h * f = f' , \qquad f'(s) = h \cdot f(s) \cdot {}^{\theta(s)}h^{-1} .$$

The orbits for this action form the set

$$H^1_\theta(G, A) = Z^1_\theta(G, A)/H .$$

We also denote by

$$H^0_\theta(G, A) = H^\theta$$

the subgroup of H consisting of the θ-invariant elements $h \in H$, i.e. such that $h = {}^{\theta(s)}h$ for any $s \in G$.

Finally we shall denote by

$$\mathbb{H}^2_\theta(G, A) \text{ and } H^2_\theta(G, A)$$

the corresponding sets endowed with the structure of not only having a "cloud" of neutral elements but also having among them a privileged one, namely the cohomology class containing the 2-cocycle $(\theta, 1)$.

The set $\mathbb{H}^2(G, A)$ is also denoted by $\underline{\underline{\text{Ext}}}^1(G, A)$.

To be able to define cohomology exact sequences we need to define the notion of a <u>short exact sequence</u> in the category $\mathcal{M}_2$ of crossed groups. This will be a diagram

(2.5) $$A' \xrightarrow{I} A \xrightarrow{J} A''$$

giving rise to the commutative diagram

$$\begin{array}{ccccccccc} & & \Pi' & \longrightarrow & \Pi & \xrightarrow{\gamma} & \Pi'' & & \\ & & {\scriptstyle\rho'}\uparrow & & {\scriptstyle\rho}\uparrow & & {\scriptstyle\rho''}\uparrow & & \\ 1 & \longrightarrow & H' & \xrightarrow[i]{} & H & \xrightarrow[j]{} & H'' & \longrightarrow & 1 \end{array}$$

satisfying the following conditions:

(i) the lower row is a short exact sequence of groups ;

(ii) $\Pi' \to \Pi$ is an isomorphism and $\gamma : \Pi \to \Pi''$ is epimorphic.

In view of condition (ii) we shall put $\Pi' = \Pi$ and replace this diagram by

(2.6) (E) $$\begin{array}{ccccccccc} & & \Pi & = & \Pi & \xrightarrow{\gamma} & \Pi'' & & \\ & & {\scriptstyle\rho'}\uparrow & I & {\scriptstyle\rho}\uparrow & J & \uparrow{\scriptstyle\rho''} & & \\ 1 & \longrightarrow & H' & \xrightarrow[i]{} & H & \xrightarrow[j]{} & H'' & \longrightarrow & 1 \end{array}$$

Let $\theta : G \to \Pi$, $\theta'' : G \to \Pi''$ be homomorphisms such that $\theta'' = \gamma \cdot \theta$. It is then possible to define maps

$$\Delta^0 = \Delta^0_E : H^0_{\theta''}(G, A'') \to H^1_\theta(G, A')$$

$$\boldsymbol{\Delta}^1 = \boldsymbol{\Delta}^1_E : \underline{\mathrm{Hom}}_{\theta''}(G, A'') \to \underline{\underline{\mathrm{Ext}}}^1(G, A')\ , \quad \Delta^1 = \Delta^1_E : H^1_{\theta''}(G, A'') \to H^2(G, A')$$

such that the following canonical square commutes

$$\begin{array}{ccc} \mathrm{Hom}_{\theta''}(G, A'') & \xrightarrow{\Delta^1} & \mathrm{Ext}^1(G, A') = \text{Ш}^2(G, A') \\ \downarrow & & \downarrow \\ H^1_{\theta''}(G, A'') & \xrightarrow[\Delta^1]{} & H^2(G, A') \end{array} .$$

Moreover

Theorem 2.2. All what we have defined is functorial and every short exact sequence (E) in $\mathcal{M}_2$ gives rise to cohomology exact sequences:

(2.7)
$$1 \to H^0_\theta(G, A') \to H^0_\theta(G, A) \to H^0_{\theta''}(G, A'') \xrightarrow{\Delta^0}$$
$$\xrightarrow{\Delta^0} H^1(G, A') \to H^1_\theta(G, A) \to H^1_{\theta''}(G, A'') \xrightarrow{\Delta^1} H^2_\theta(G, A') \to H^2_\theta(G, A) \to H^2_{\theta''}(G, A''),$$

(2.8)
$$* \to \underline{\mathrm{Hom}}_\theta(G, A') \to \underline{\mathrm{Hom}}_\theta(G, A) \to \underline{\mathrm{Hom}}_{\theta''}(G, A'') \xrightarrow{\Delta^1}$$
$$\xrightarrow{\Delta^1} \underline{\underline{\mathrm{Ext}}}^1_\theta(G, A') \to \underline{\underline{\mathrm{Ext}}}^1_\theta(G, A) \to \underline{\underline{\mathrm{Ext}}}^1_{\theta''}(G, A'') .$$

In this statement, exactness means that the sets in these sequences are endowed with structures which make it possible to answer the following two questions:

Problem (a). When is an element in one set (except the last one) the image of the preceding arrow?

Problem (b). When do two elements in one set have the same image through the next arrow?

For example each set is pointed or has a cloud of neutral "droplets" with a more distinguished one. This allows to solve question (a). But the structure necessary to solve question (b) is more complicated and more interesting. For example each

Hom$_\theta$(G, A) is a polypus and each $\underline{\underline{\mathrm{Ext}}}^1$(G, A) = $\mathbb{H}^2$(G, A) is sitting under an object which I would like to call a lobster, namely a bigger animal which has pinces.

3. Approach of three-dimensional cohomology.

This aims at solving question (a) for the last object in the sequences (2.6) and (2.7).

To be more geometric let us remind that any $\xi \in \underline{\underline{\mathrm{Ext}}}^1_\theta(G, A)$ can be represented by an isomorphism class of diagrams

(3.1)
$$\begin{array}{ccccccccc} & & & & \Pi & & & & \\ & & & & \uparrow{\scriptstyle \sigma} & & & & \\ 1 & \longrightarrow & H & \underset{i}{\longrightarrow} & X & \longrightarrow & G & \longrightarrow & 1 \end{array}$$

with the row an exact sequence of groups and σ a homomorphism subject to conditions

(i) $\rho = \sigma \cdot i$,

(ii) for $x \in X$, $h \in H$, $i({}^{\sigma x}h) = x \cdot ih \cdot x^{-1}$.

These diagrams are functorial in the sense the a morphisms (j, γ) in $\mathcal{M}_2$ as in (2.1) produces a sort of push-out diagram

(3.2)
$$\begin{array}{ccccccccc} & & & & & \Pi & & & \\ & & & & {\scriptstyle \sigma}\nearrow & \downarrow{\scriptstyle \gamma} & & & \\ 1 & \longrightarrow & H & \longrightarrow & X & \longrightarrow & G & \longrightarrow & 1 \\ & & \downarrow{\scriptstyle j} & & \downarrow & \Pi' & \Vert & & \\ & & & & & {\scriptstyle \sigma'}\nearrow & & & \\ 1 & \longrightarrow & H' & \longrightarrow & X' & \longrightarrow & G & \longrightarrow & 1 \end{array}$$

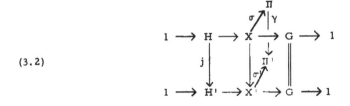

unique up to isomorphism. In particular take $H' = H$, $\Pi' = \Pi$, $\gamma = id$, and let $j = j^{\alpha} : H \to H$ be the automorphism induced by some $\alpha \in \Pi$. Then (3.2) can be considered as an isomorphism between two diagrams of type (3.1). These larger isomorphism classes can be identified with the elements of the thin $H^2(G, A)$.

Problem (a) for the last object in the sequence (2.7) is equivalent to: supposing that in (2.1) j and γ are epimorphisms and we have in (3.2) the bottom part, namely

$$\begin{array}{ccccccccc} & & & & \Pi' & & & & \\ & & & & \uparrow{\scriptstyle\sigma'} & & & & \\ 1 & \longrightarrow & H' & \longrightarrow & X' & \longrightarrow & G & \longrightarrow & 1 \end{array} ,$$

is it possible to find some diagram (3.1) fitting into (3.2)? And similarly for the last object of sequence (2.6).

Let us now go back to the exact sequence (2.4) in $\mathcal{M}_2$ and suppose that $\Lambda \subset \Pi$ is the kernel of $\gamma : \Pi \to \Pi''$, so that we have the exact sequence

$$1 \longrightarrow \Lambda \longrightarrow \Pi \longrightarrow \Pi'' \longrightarrow 1 .$$

Next, let us consider a 2-cocycle $(\phi'', h'') \in Z^2(G, A'')$. It is then certainly possible to find a 2-cochain (ϕ, h) with values in A, namely $(\phi, h) \in C^2(G, A)$ which lifts (ϕ'', h'') in the sense that $J_{\#}(\phi, h) = (\phi'', h'')$. We then have functions

$$\lambda : G \times G \longrightarrow \Lambda , \quad k : G \times G \times G \longrightarrow H'$$

such that

$$\phi(s)\,\phi(t) = \lambda(s,t)\,\rho h(s,t)\,\phi(st) \tag{3.3}$$

$$^{\phi(s)}h(t,u) \cdot h(s,tu) = k(s,t,u) \cdot h(s,t) \cdot h(st,u) . \tag{3.4}$$

It thus seems that the pair (k,λ) is a candidate to the title of 3-cochain and ultimately of 3-cocycle. However another lifting (ϕ',h') of (ϕ'',h'') is obtained by applying to (ϕ,h) a deviation (α,a) such that

$$(\phi',h') = (\alpha,a)\cdot(\phi,h) = (\alpha\cdot\phi,\ a\cdot h)$$

where

$$\alpha : G \to \Lambda \qquad \text{and} \qquad a : G\times G \to H'$$

are arbitrary functions. The new lifting (ϕ',h') then produces a pair (k',λ') and one would like to compute it as a function of the initial (k,λ) and the deviation (α,a).

Unfortunately such a formula does not exist and easy calculations yield:

$$\lambda'(s,t) = \alpha(s)\cdot{}^{\phi(s)}\alpha(t)\cdot\lambda(s,t)\cdot{}^{\rho h(s,t)}\alpha(st)^{-1}\cdot\rho a(s,t)^{-1},$$

$$k'(s,t,u) = {}^{\alpha(s)\phi(s)}a(t,u)\cdot{}^{\phi(s)}_{\alpha(s)}L(t,u)\cdot[{}^{\phi(s)}\rho h(t,u)]_{a(s,tu)}\cdot k(s,t,u)\cdot{}^{\rho h(s,t)}a(st,u)^{-1}\cdot a(s,t)^{-1}.$$

The real trouble is the presence of the expression

$${}^{\phi(s)}_{\alpha(s)}L(t,u) = {}^{\alpha(s)\phi(s)}h(t,u)\cdot{}^{\phi(s)}h(t,u)^{-1}$$

and we reach here the <u>first crucial problem</u> in the construction of 3-cohomology. Moreover, in both formulas appear the function ϕ which means that we have to add it to the pair (k,λ). But we also have to incorporate a function

$$L : \Lambda\times\Pi\times G\times G \to H'$$

the value of which at (α,ϕ,s,t) we want to denote

$$L(\alpha,\phi,s,t) = {}^{\phi}_{\alpha}L(s,t) = {}^{\phi}_{\alpha}L_{s,t}$$

and which is given by

$$ {}^{\phi}_{\alpha}L(s,t) = {}^{\alpha\cdot\phi}h(s,t)\cdot {}^{\phi}h(s,t)^{-1} . $$

This is actually a nice function since it measures the modification undergone by ${}^{\phi}h(s,t) \in H$, when twisted by $\alpha \in \Lambda$, this modification being by necessity an element of the smaller group H'. Also, for ϕ, s, t fixed, this function is an <u>inverse crossed homomorphism</u> $L : \Lambda \to H'$ (namely its inverse is a crossed homomorphism $\Lambda \to H'$ or one has

$$ {}_{\alpha\beta}L = {}^{\alpha}({}_{\beta}L) \cdot {}_{\alpha}L) . $$

We can however consider the more simple function

(3.5) $$ K : \Lambda \times G \times G \to H' $$

defined by

(3.6) $$ K(\alpha, s, t) = {}_{\alpha}K(s,t) = {}^{\alpha}h(s,t) \cdot h(s,t)^{-1} $$

which measures the non invariantness of $h : G \times G \to H$ by $\alpha \in \Lambda$. We then have

$$ {}^{\phi}_{\alpha}L(s,t) = {}^{\phi}[{}_{\phi^{-1}\alpha\phi}K] . $$

This can be clarified as follows. Consider the set

$$ C^1 = C^1(\Lambda, H') = \underline{\mathrm{App}}(\Lambda, H') $$

and define a group action $*$ of Π onto C^1 (composition product) such that for $\phi \in \Pi$ and $K \in C^1$, $\phi * K : \Lambda \to H'$ is defined by

$$ (\phi * K)(\alpha) = {}^{\phi}[K(\phi^{-1}\alpha\phi)] $$

or equivalently

$$^{\phi}[K(\alpha)] = (\phi * K)(^{\phi}\alpha) , \qquad ^{\phi}\alpha = \phi\alpha\phi^{-1} .$$

Similarly, if K is as in (3.5), it can be identified with a map $K: G \times G \to C^1$ and then we get $\phi * K: G \times G \to C^1$. With this formalism it is clear that

(3.7) $$^{\phi}_{\alpha}L(s,t) = {}_{\alpha}(\phi * K)(s,t) .$$

To simplify the notations we shall also write $\phi * K = {}^{\phi}K$ and

(3.8) $$^{\phi}_{\alpha}L(s,t) = {}^{\phi}_{\alpha}K(s,t) .$$

We then want to consider a 3-cochain as a system $(k, \lambda, \phi; K, \eta)$ of five functions

$$k: G \times G \times G \to H' ,$$

$$\lambda: G \times G \to \Lambda ,$$

$$\phi: G \to \Pi ,$$

$$K: \Lambda \times G \times G \to H' ,$$

$$\eta: G \times G \to \Pi .$$

Now, starting with a lifting $(\phi, h) \in C^2(G, A)$ of $(\phi'', h'') \in Z^2(G, A'')$, we have obtained such a system [with $\eta(s,t) = \rho h(s,t)$] which we shall denote by

(3.9) $$(k, \lambda, \phi; K; \eta) = \square (\phi, h) .$$

Thus $\square$ represents sort of a coboundary operator

$$\square : C^2(G, A) \to C^3(G, ?) .$$

At this point the lector can easily check that the pairs (α, a) form a group $C^2(G, B)$ where B is the crossed group $(H', \rho', \Lambda, \Psi)$ in which Ψ is the action of Λ to H' restricted from the one Φ' of Π onto H'. Moreover there is a group action

$$C^2(G, B) \times C^2(G, A) \to C^2(G, A),$$

$$((\alpha, a), (\phi, h)) \rightsquigarrow (\alpha, a) \cdot (\phi, h) = (\phi', h') .$$

We also have a group action

$$* : C^2(G, B) \times C^3(G, ?) \to C^3(G, ?) \tag{3.10}$$

$$((\alpha, a), (k, \lambda, \phi; K; \eta)) \rightsquigarrow (k', \lambda', \phi'; K'; \eta')$$

such that

$$\square [(\alpha, a) \cdot (\phi, h)] = (\alpha, a) * [\square (\phi, h)] . \tag{3.11}$$

This action is explicitly defined by putting

$$k'(s,t,u) = {}^{\alpha(s)\phi(s)}a(t,u) \cdot {}^{\phi(s)}_{\alpha(s)}K(t,u) \cdot [{}^{\phi(s)}\eta(t,u)]_{a(s,tu)} \cdot k(s,t,u) \cdot \tag{3.12}$$

$$\cdot {}^{\eta(s,t)}a(st,u)^{-1} \cdot a(s,t)^{-1} ,$$

$$\lambda'(s,t) = \alpha(s) \cdot {}^{\phi(s)}\alpha(t) \cdot \lambda(s,t) \cdot {}^{\eta(s,t)}\alpha(st)^{-1} \cdot \rho a(s,t)^{-1} , \tag{3.13}$$

$$\phi'(s) = \alpha(s) \cdot \phi(s) , \tag{3.14}$$

$$_{\mu}K'(s,t) = {}^{\mu}a(s,t) \cdot {}_{\mu}K(s,t) \cdot a(s,t)^{-1} , \tag{3.15}$$

$$\eta'(s,t) = \rho a(s,t) \cdot \eta(s,t) . \tag{3.16}$$

The existence of the action (3.10) giving rise to formula (3.11) shows that the 3-cochain $(k', \lambda', \phi'; K'; \eta')$ derived from (ϕ', h') can be computed in terms of $(k, \lambda, \phi; K; \eta)$ and the deviation (α, a). Thus the 5-uples $(k, \lambda, \phi; K; \eta)$ are really sensible candidates to the titles of 3-cochains and 3-cocycles.

Next we have to find out identities satisfied by the 3-cochains defined by (3.9). These identities will define a subset

$$Z^3(G, ?) \subset C^3(G, ?)$$

of 3-cocycles and we shall have to know what ? means, namely what the system of coefficients is.

We reach here the <u>second crucial problem</u>, namely deriving an identity on the component $k: G \times G \times G \to H'$. This identity should generalize the well known condition

(3.17) $$(\delta k)(s,t,u,v) = {}^{\phi(s)}k(t,u,v) - k(st,u,v) + k(s,tu,v) - k(s,t,uv) - k(s,t,u) = 0$$

of the classical theory. This we do as follows. We have

(3.18) $$k(s,t,u) = {}^{\phi(s)}h(t,u) \cdot h(s,tu) \cdot h(st,u)^{-1} . h(s,t)^{-1}$$

together with

(3.19) $${}^{\phi(s)}h(t,u) = k(s,t,u) \cdot h(s,t) \cdot h(st,u) \cdot h(s,tu)^{-1} .$$

We then use (3.18) to write

$${}^{\phi(s)}k(t,u,v) = {}^{\phi(s)\phi(t)}h(u,v) . {}^{\phi(s)}h(t,uv) . {}^{\phi(s)}h(tu,v)^{-1} . {}^{\phi(s)}h(t,u)^{-1} .$$

This is then transformed using (3.3) and four different forms of (3.19) to yield

$$(3.20.\mathrm{I}) \qquad {}^{\phi(s)}k(t,u,v)\cdot k(s,t,u)\cdot {}^{\rho H(s,t,u)}k(s,tu,v) =$$

$$\lambda(s,t)\cdot {}^{\eta(s,t)}k(st,u,v)\cdot N(s,t,u,v)\cdot [{}^{\eta(s,t)}\rho H(st,u,v)]_{k(s,t,uv)}\ .$$

In there one has

$$(3.21) \qquad H(s,t,u) = h(s,t)\cdot h(st,u)\cdot h(s,tu)^{-1}\ ,$$

$$(3.22) \qquad N(s,t,u,v) = {}^{\lambda(s,t)\cdot\eta(s,t)}H(st,u,v)\cdot {}^{\eta(s,t)}H(st,u,v)^{-1}\ .$$

Moreover $\eta(s,t)$ represents $\rho h(s,t)$ and can be computed in terms of the components λ, ϕ by the formula

$$(3.20.\mathrm{II}) \qquad \phi(s)\,\phi(t) = \lambda(s,t)\cdot\eta(s,t)\cdot\phi(st)\ .$$

Last but not least the function N can be computed out of the function K, so that (3.20.I) is actually an identity on the 5-uple $(k, \lambda, \phi; K; \eta)$. The assiduous lector will indeed check that

$$N(s,t,u,v) = {}_{\lambda}K\cdot {}^{\eta}({}_{\lambda}K_1)\cdot {}^{\eta\eta_1}({}_{\lambda}K_2)^{\eta\eta_1\eta_2\eta_3^{-1}}({}_{\lambda}K_3^{-1})\cdot {}^{\eta\eta_1\eta_2\eta_3^{-1}\eta^{-1}}({}_{\lambda}K^{-1})$$

with

$${}_{\lambda}K = {}^{\lambda(s,t)}K(s,t)\ , \qquad {}_{\lambda}K_1 = {}^{\lambda(s,t)}K(st,u)$$

$${}_{\lambda}K_2 = {}^{\lambda(s,t)}K(stu,v) \qquad {}_{\lambda}K_3 = {}^{\lambda(s,t)}K(st,uv)\ .$$

<u>Remark</u> 3.1. It is useful to observe that formula (3.20.I) obviously has the shape of (3.17) and it also has the shape of the diagram

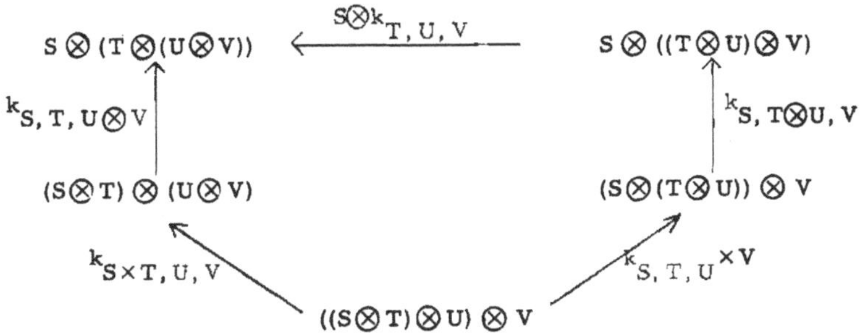

the commutativity of which is a relation among the canonical isomorphisms

$$k_{S,T,U} : (S \otimes T) \otimes U \rightarrow S \otimes (T \otimes U)$$

expressing the defect of associativity of the tensor product of modules. The only snag is that the first member of (3.20.I) should be written as

$$^{\phi(s)}k(t,u,v) \cdot {}^{\rho k(s,t,u) \cdot \rho H(s,t,u)}k(s,tu,v) \cdot k(s,t,u) \; .$$

But precisely the function k is an obstruction to the associativity of a product defined on the set $H \times G$ by

$$(m.s) \cdot (n,t) = (m.{}^{\phi(s)}n \cdot h(s,t), st)$$

where $m, n \in H$, $s, t \in G$.

<u>Remark</u> 3.2. Let h be an element of H and a an element of Λ. We then define the function $\partial h : \Lambda \rightarrow H'$, putting

$$(\partial h)(a) = \partial_a h = {}^a h \cdot h^{-1} \; .$$

The same formalism applies when h is a function from some set E to H; then ∂h is a function from $\Lambda \times E$ to H' (e.g. the function K of (3.6) is just ∂h for

$h: G \times G \to H$). We have also the function $\delta h : \Lambda \to H'$

$$(\delta h)(\alpha) = \delta_\alpha h = h \cdot {}^\alpha h^{-1} .$$

We call the operators δ and ∂ the <u>coboundary</u> and <u>antiboundary</u> operators. δh is a crossed homomorphism while $\partial h = (\delta h)^{-1}$ is an inverse crossed homomorphism. Considering that the group Π operates onto H by automorphisms and onto $C^1(\Lambda, H')$ by the composition product $*$, it is clear that ∂ and δ are equivariant, namely

$$\partial({}^\phi h) = \phi * (\partial h), \qquad \delta({}^\phi h) = \phi * (\delta h) .$$

We can also consider ∂h as a function in between the larger groups Π and H and then observe that, in this frame,

$$\partial_\alpha({}^\phi h) = {}^{\alpha\phi}h \cdot {}^\phi h^{-1} = ({}^{\alpha\phi}h \cdot h^{-1}) \cdot ({}^\phi h \cdot h^{-1})^{-1} = (\partial_{\alpha\phi} h) \cdot (\partial_\phi h)^{-1} .$$

More generally, if $K : \Lambda \to H'$ is an inverse crossed homomorphism, one has the formula (justifying the writing ${}_\alpha K$ instead of $K(\alpha)$)

(3.23) $$ {}_\alpha(\beta * K) = {}_{\alpha\beta}K \cdot {}_\beta K^{-1} .$$

This can be seen as follows: K verifies

$$ {}_{\alpha\beta}K = {}^\alpha({}_\beta K) \cdot {}_\alpha K$$

so that

(3.24) $$({}_\alpha K)^{-1} = {}^\alpha({}_{\alpha^{-1}}K) \quad \text{or} \quad K(\alpha)^{-1} = {}^\alpha(K(\alpha^{-1})) .$$

Then one proves (3.23) by writing

$$ {}_\alpha(\beta * K) = {}^\beta({}_{\beta^{-1}\alpha\beta}K) = {}^\beta[{}^{\beta^{-1}}({}_{\alpha\beta}K) \cdot {}_{\beta^{-1}}K] = {}_{\alpha\beta}K \cdot ({}_\beta K)^{-1} .$$

According to previous notations, the set of crossed homomorphisms from Λ to H' is denoted $Z^1(\Lambda, H') = Z^1_{\Psi}(\Lambda, H')$. (Remind that Ψ is the action of Λ onto H' in the crossed group B of (3.10).) We shall denote by

$$S^1(\Lambda, H') = S^1_{\Psi}(\Lambda, H')$$

the set of inverse crossed homomorphisms.

<u>Remark</u> 3.3. We also want to show that if $K : \Lambda \to H'$ is an inverse crossed homomorphism, so is also $\phi * K$ for any $\phi \in \Pi$. Indeed

$$(\phi * K)(\alpha\beta) = {}^{\phi}[K(\phi^{-1}\alpha\beta\phi)] = {}^{\phi}[{}^{\phi^{-1}\alpha\phi}K(\phi^{-1}\beta\phi)\cdot K(\phi^{-1}\alpha\phi)]$$

$$= {}^{\alpha\phi}K(\phi^{-1}\beta\phi)\cdot {}^{\phi}K(\phi^{-1}\alpha\phi) = {}^{\alpha}[(\phi * K)(\beta)]\cdot(\phi * K)(\alpha) .$$

We now derive three more identities on the 5-uple $(k, \lambda, \phi; K; \eta) = \square(\phi, h)$.

One is just the fact that K is an inverse crossed homomorphism, namely

(3.20.III) $$_{\alpha\beta}K = {}^{\alpha}({}_{\beta}K)\cdot {}_{\alpha}K \quad \text{or} \quad {}_{\alpha\beta}K(s,t) = {}^{\alpha}[{}_{\beta}K(s,t)]\cdot {}_{\alpha}K(s,t) .$$

To obtain the next identity we twist formula (3.4) by α to get

$${}^{\alpha\phi(s)}h(t,u)\cdot_{\blacklozenge}{}^{\alpha}h(s,tu) = {}^{\alpha}k(s,t,u)\cdot{}^{\alpha}h(s,t)\cdot{}^{\alpha}h(st,u) .$$

Then inserting in the spot $\blacklozenge$ the trival expression

$${}^{\phi(s)}h(t,u)^{-1}\cdot k(s,t,u)\cdot h(s,t)\cdot h(st,u)\cdot h(s,tu)^{-1}$$

yields

$${}^{\phi(s)}{}_{\alpha}L(t,u)\cdot k(s,t,u)\cdot h(s,t)\cdot h(st,u)\cdot h(s,tu)^{-1}\cdot{}^{\alpha}h(s,tu) = {}^{\alpha}k(s,t,u)\cdot{}^{\alpha}h(s,t)\cdot{}^{\alpha}h(st,u)$$

and finally, putting $\Theta(s,t,u) = \rho H(s,t,u)$:

$$(3.20.\mathrm{IV}) \qquad {}^{\phi(s)}_{\ a}K(t,u)\cdot k(s,t,u)\cdot {}^{\Theta(s,t,u)}[{}_aK(s,tu)] = {}^{a}k(s,t,u)\cdot {}_aK(s,t)\cdot {}^{\eta(s,t)}{}_aK(st,u).$$

The analogy between this formula and the fundamental identity (2.2.a) on 2-cocycles is clear.

We finally have the following identity, which plays a role similar to the second identity (2.2.b) for 2-cocycles:

$$(3.20.\mathrm{V}) \qquad {}^{\phi(s)}\lambda(t,u)\cdot {}^{[{}^{\phi(s)}\eta(t,u)]}\lambda(s,tu)\cdot \rho k(s,t,u) = \lambda(s,t)\cdot {}^{\eta(s,t)}\lambda(st,u).$$

We derive it from (3.3), computing in two different ways the product $\phi(s)\cdot\phi(t)\cdot\phi(u)$. When a and β are elements of Π, we have written

$${}^{a}\beta = a\beta a^{-1} .$$

4. Definition of three-dimensional cohomology

We have now obtained good candidates for 3-cochains, namely the 5-uples of functions $(k, \lambda, \phi; K; \eta)$ as above. Among them, those satisfying the conditions (3.20.I-II-III-IV-V) are candidate 3-cocycles. However so far the coefficient system is not yet clear. A candidate would be the system

$$\mathbf{A}' = (H', \rho', \Pi, \Phi'; \Lambda)$$

in which $A' = (H', \rho', \Pi, \Phi')$ is a crossed group and in which Λ is a normal subgroup of Π such that

$$\rho' H' \subset \Lambda \subset \Pi .$$

However we shall face more difficulty when trying to define a connecting map and a longer exact sequence. This will oblige us to incorporate more structure in the coefficient system.

Going back to a 2-cocycle $(\phi'', h'') \in Z^2(G, A'')$, we have lifted it into $(\phi, h) \in C^2(G, A)$ and formed

$$\square(\phi, h) = (k, \lambda, \phi; K; \eta) \in Z^3(G, A') .$$

However (ϕ, h) is defined up to a pair $(\alpha, a) \in C^2(G, B)$ (where $B = (H', \rho', \Lambda, \Psi)$ is the crossed group appearing in (3.10)). Replacing (ϕ, h) by $(\phi', h') = (\alpha \cdot \phi, a \cdot h)$ yields a system $(k', \lambda', \phi'; K'; \eta')$ related to the preceding formulas (3.12) to (3.16) which precisely represent the action (3.10). They also represent what we shall call the <u>first variation</u> of the 3-cocycle $(k, \lambda, \phi; K; \eta)$.

The <u>second variation</u> shall be obtained by moving the 2-cocycle (ϕ'', h'') along its thick cohomology class, namely applying to it the action (2.3). To that effect we consider $b'' : G \to H''$ and produce $b'' * (\phi'', h'') = (\overline{\phi}'', \overline{h}'')$. Then b can be lifted into some function $b : G \to H$ and the lifting (ϕ, h) of (ϕ'', h'') is then transformed into $b * (\phi, h) = (\overline{\phi}, \overline{h})$ which is a lifting of $(\overline{\phi}'', \overline{h}'')$. The problem is then to compute $\square(\overline{\phi}, \overline{h}) = (\overline{k}, \overline{\lambda}, \overline{\phi}; \overline{K}; \overline{\eta})$ in terms of $\square(\phi, h) = (k, \lambda, \phi; K; \eta)$ and some operation to be described inside of the coefficient system. However it will turn out that such an operation cannot be described in terms of our candidate A' : this will lead to the <u>third crucial step</u> in the development of 3-cohomology.

Easy computations yield the following

$$\overline{k}(s,t,u) = b(s) . {}^{\phi(s)}b(t) . {}^{\phi(s)\phi(t)}b(u) \cdot k(s,t,u) . {}^{\eta(s,t) \cdot \phi(st)}b(u)^{-1} . {}^{\phi(s)}b(t)^{-1} . b(s)^{-1}$$

or, putting $\beta(s) = \rho\, b(t)$,

$$(4.1)\qquad \bar{k}(s,t,u) = \beta(s)\,.\,{}^{\phi(s)}\beta(t)\,.\,{}^{\phi(s)\phi(t)}\beta(u)\,k(s,t,u)\,.\,\beta(s)\,.\,{}^{\phi(s)}\beta(t)\,[\partial_{\lambda(s,t)^{-1}}{}^{\phi(s)\phi(t)}b(u)]^{-1}\;;$$

$$(4.2)\qquad \bar{\lambda}(s,t) = \beta(s)\,.\,{}^{\phi(s)}\beta(t)\,\lambda(s,t)\;;$$

$$(4.3)\qquad \bar{\phi}(s) = \beta(s)\cdot\phi(s)\;;$$

$$(4.4)\qquad {}_a\bar{K}(s,t) = [\partial_a b(s)]\,.\,{}^{\beta(s)}[\partial_a{}^{\phi(s)}b(t)]\,.\,{}^{\beta(s)\,.\,{}^{\phi(s)}\beta(t)}[{}_aK(s,t)]\,.$$

$$.\,{}^{\beta(s)\,.\,{}^{\phi(s)}\beta(t)\cdot\eta(s,t)\cdot\beta(st)^{-1}}[\partial_a b(st)]^{-1}\;.$$

$$(4.5)\qquad \bar{\eta}(s,t) = \beta(s)\,.\,{}^{\phi(s)}\beta(t)\cdot\eta(s,t)\cdot\beta(st)^{-1}\;.$$

In these formulas we have set

$$(4.6)\qquad \partial_a{}^{\phi}h = \partial_a({}^{\phi}h) = {}_a[\partial({}^{\phi}h)] = {}_a[\,\phi*(\partial h)]\;.$$

There is finally a <u>third variation</u> if we allow the 2-cocycle (ϕ'', h'') to move into its thin cohomology class, specifically if we move it under the action (2.6) of Π''. This however produces the very sweet formulas (4.7) to (4.11). Suppose (ϕ'', h'') is brought by $a'' \in \Pi''$ into $(\tilde{\phi}'', \tilde{h}'')$:

$$\tilde{\phi}''(s) = {}^{a''}\phi''(s), \qquad \tilde{h}''(s,t) = {}^{a''}h(s,t)\;.$$

Lifting $a'' \in \Pi''$ by $\pi \in \Pi$ brings the lifting (ϕ, h) into a lifting $(\tilde{\phi}, \tilde{h})$ of $(\tilde{\phi}'', \tilde{h}'')$ such that

$$\tilde{\phi}(s) = {}^{\pi}\phi(s) = \pi\cdot\phi(s)\cdot\pi^{-1}\,, \qquad \tilde{h}(s,t) = {}^{\pi}h(s,t)\;.$$

Then $\square(\tilde{\phi}, \tilde{h}) = (\tilde{k}, \tilde{\lambda}, \tilde{\phi}; \tilde{K}; \tilde{\eta})$ is given by

(4.7) $$\widetilde{k}(s,t,u) = {}^{\pi}k(s,t,u) ,$$

(4.8) $$\widetilde{\lambda}(s,t) = {}^{\pi}\lambda(s,t) ,$$

(4.9) $$\widetilde{\phi}(s) = {}^{\pi}\phi(s) ,$$

(4.10) $${}_{a}\widetilde{K}(s,t) = {}_{a}(\pi^* K)(s,t) \text{ or } \widetilde{K} = \pi^* K ,$$

(4.11) $$\widetilde{\eta} = {}^{\pi}\eta = \pi\eta\pi^{-1} .$$

Let us now go back to the second variation.

We would have liked to eliminate the group H (which does not appear in our candidate coefficient) out of formulas (4.1) to (4.5). Unfortunately this is not possible because we have to know the expression ∂b with $b : G \to H$. The presence of $\rho b = \beta$ is of course no trouble since it lies into $\mathbf{A}'$.

To go around this difficulty we ask ourselves under what condition two elements b and b' of H produce at the same time

$$\rho b = \rho b' \quad \text{and} \quad \partial b = \partial b' .$$

Suppose $b' = b \cdot x$, $x \in H$. Then one should have

$$\rho x = 1 \text{ or } x \in Z_H = \underline{\text{Ker}}\ \rho, \text{ the center of A}$$

and

$$\text{for every } a \in \Lambda, \quad {}^{a}x = x .$$

The second condition means that x belongs to the Π-invariant subgroup of Λ-invariant elements of H. Thus the intersection of this group and Z_H is a normal and even Π-invariant subgroup P of H. We thus want to consider the quotient group $E = H/P$ and the quotient map $\kappa : H \to E$ fitting into the diagram

(4.12)

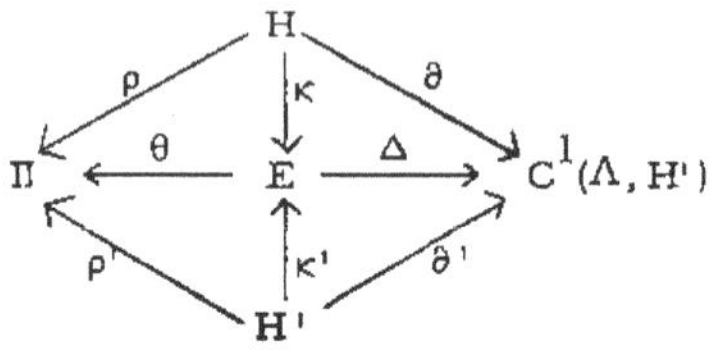

which defines uniquely θ and Δ and in which $\kappa' = \kappa i$, $\partial' = \partial i$, $i : H' \to H$. There exists also a canonical action of Π onto E such that κ is equivariant as well as Δ in the sense that

$$\Delta({}^{\phi}m) = \phi^* \Delta m , \qquad m \in E , \ \phi \in \Pi .$$

Suppose now we have a map $b : G \to H$ producing $m = \kappa \cdot b : G \to E$. We can then transliterate the delicate formulas (4.1) and (4.4) as follows (we put $\beta(s) = \theta\, m(s)$ and use a convention similar to (4.6)):

$$(4.1') \qquad \overline{k}(s,t,u) = \beta(s) . {}^{\phi(s)}\beta(t) . {}^{\phi(s)\phi(t)}\beta(u)_{k(s,t,u)} . \beta(s) . {}^{\phi(s)}\beta(t)[\Delta^{\phi(s)\phi(t)}_{\lambda(s,t)^{-1}} m(u)]^{-1} ;$$

$$(4.4') \qquad {}_a\overline{K}(s,t) = [\Delta_a m(s)] . {}^{\beta(s)}[\Delta_a^{\phi(s)} m(s)] . {}^{\beta(s) . {}^{\phi(s)}\beta(t)}[{}_a K(s,t)] .$$

$$. \, {}^{\beta(s) . {}^{\phi(s)}\beta(t) \cdot \eta(s,t) . \beta(st)^{-1}}[\Delta_a m(st)]^{-1} .$$

<u>Definition</u> 4.1. By a <u>super-crossed</u> <u>group</u> we mean a system

$$\mathbb{A} = (H, \rho, \Pi, \Phi ; \Lambda; E, \Psi , \kappa, \theta, \Delta)$$

as follows:

(1°) Λ is a normal subgroup of Π such that

$$\rho H \subset \Lambda \subset \Pi ;$$

(2°) $A = (H, \rho, \Pi, \Phi)$ is a crossed group and thus $B = (H, \rho, \Lambda, \Phi)$ as well;

(3°) E is a group endowed with an action Ψ of Π ;

(4°) $\kappa : H \to E$, $\theta : E \to \Pi$, $\Delta : E \to C^1(\Lambda, H)$ are equivariant homomorphisms fitting into a commutative diagram

$$\begin{array}{ccccc} & & H & & \\ & \swarrow^{\rho} & \downarrow{\scriptstyle\kappa} & \searrow^{\partial} & \\ \Pi & \xleftarrow{\theta} & E & \xrightarrow{\Delta} & S^1(\Lambda, H) ; \end{array}$$

(5°) $C = (E, \theta, \Pi, \Psi)$ is a crossed group and $\kappa : H \to E$ induces a morphism of A to C in $\mathfrak{M}_2$.

(6°) For $m \in E$, $a \in \Lambda$, $\rho(\Delta_a m)$ is the commutator $[a, \theta m] = a . \theta m . a^{-1} . \theta m^{-1}$.

If $\mathbf{A}' = (H', \rho', \Pi'; \Lambda'; E', \Psi', \kappa', \theta' \Delta')$ is another super-crossed group, a morphism $\mathbf{A} \to \mathbf{A}'$ will consist of a system of morphisms

$$H \to H', \quad \Pi \to \Pi', \quad \Lambda \to \Lambda', \quad E \to E'$$

subject to conditions which we leave to the lector. We thus are able to speak of the category $\mathfrak{M}_3$ of super-crossed groups.

Example 4.2.1. Consider a crossed group $A = (H, \rho, \Pi, \phi)$ together with a group Λ satisfying condition (2°). This allows to construct E, Ψ and the upper part of diagram (4.12). Then

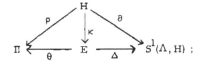

$$\mathbb{A} = (H, \rho, \Pi, \phi; \Lambda; E, \Psi, \kappa, \theta, \Delta)$$

is an example of a supercrossed group. We could take $\Lambda = \rho H$ and would then call $\mathbb{A}$ the supercrossed group associated to the crossed group A.

Example 4.2.2. Consider a short exact sequence (E) of crossed-modules as in (2.4). Then with the above notations

$$\mathbb{A}' = (H', \rho', \Pi, \Phi; \Lambda; E, \Psi, \kappa', \theta, \Delta)$$

is another example. We call this supercrossed group the first supercrossed group associated to (E).

Definition 4.3. By a 3-cochain of G with values into $\mathbb{A}$, we mean a system $(k, \lambda, \phi; K; \eta)$ of functions

$$k : G \times G \times G \to H,$$
$$\lambda : G \times G \to \Lambda,$$
$$\phi : G \to \Pi,$$
$$K : G \times G \to C^1(\Lambda, H) \quad \text{or} \quad K : \Lambda \times G \times G \to H,$$
$$\eta : G \times G \to \Pi.$$

This 3-cochain is called a 3-cocycle if and only if it satisfies conditions (3.20.I to V). We denote by $Z^3(G, \mathbb{A})$ the set of these 3-cocycles. A 3-cocycle $(k, \lambda, \phi; K; \eta)$ is said to be neutral if its components k and λ are the trivial maps with a single value, the units in the groups H and Λ respectively.

We want to consider three actions onto $Z^3(G, A)$. The first one * corresponds to the first variation and is the one described in (3.10) and formulas (3.12) to (3.16), the acting group being $C^2(G,B)$. The next one corresponds to the second variation and has $C^1(G,E)$ as operating group; it shall be denoted

$$\nabla : C^1(G,E) \times Z^3(G,\mathbb{A}) \to Z^3(G,\mathbb{A}) .$$

It is described by formulas analogous to (4.1 '), (4.2), (4.3),(4.4'), (4.5). Finally we allow for a third action by the group Π, corresponding to the third variation, described by formulas (4.7) to (4.11).

The orbits ω of $Z^3(G,A)$ corresponding to the first action form a set which could be called a <u>super-thick</u> 3-<u>cohomology</u> $\mathbf{H}^3(G,\mathbb{A})$ but is not very interesting. It turns out that the second action operates also on the set of orbits ω. In other words an element $m \in C^1(G,E)$ transforms an orbit ω in an other one $\omega' = m\cdot\omega$. The union of these orbits $m\cdot\omega$ is then a bigger orbit Ω for a bigger action. The set of orbits Ω then forms what we shall call the <u>thick</u> 3-<u>cohomology set</u> $\mathbb{H}^3(G,\mathbb{A})$ endowed with a canonical projection

$$\gamma : \mathbf{H}^3(G,\mathbb{A}) \to \mathbb{H}^3(G,\mathbb{A}) .$$

Finally, it turns out that the third action operates on the set of orbits Ω. In other words an element $\pi \in \Pi$ transforms an orbit Ω into another one $\Omega' = \pi\Omega$. The union of these orbits $\pi\cdot\Omega$ forms then a still bigger orbit $\mathbf{\Omega}$ for a bigger action. The set of these big orbits $\mathbf{\Omega}$ constitutes then a set which we shall call the <u>thin</u> 3-<u>cohomology set</u> $H^3(G,\mathbb{A})$ and is endowed with a canonical projection

$$\gamma : \mathbb{H}^3(G,\mathbb{A}) \to H^3(G,\mathbb{A}) .$$

In order to make clear the statement that an element $m \in C^1(G,E)$ transforms an orbit ω into another one of the same type, let us mention the following. Suppose ω is the orbit of $\underline{\underline{k}} = (k,\lambda,\phi;K;\eta)$. Then moving $\underline{\underline{k}}$ by $(\alpha,a) \in C^2(G,B)$ into $\underline{\underline{k}}' = (\alpha,a) * \underline{\underline{k}}$ and then moving $\underline{\underline{k}}'$ by $m \in C^1(G,E)$ into $\overline{\underline{\underline{k}}}' = m \nabla \underline{\underline{k}}'$ is equivalent to the following: moving $\underline{\underline{k}}$ by m into $\overline{\underline{\underline{k}}} = m \nabla \underline{\underline{k}}$ and then moving $\overline{\underline{\underline{k}}}$ by some $(\bar{\alpha},\bar{a}) \in C^2(G,B)$ into $(\bar{\alpha},\bar{a}) * \overline{\underline{\underline{k}}}$ where

$$\bar{\alpha}(s) = \beta(s)\,\alpha(s)\,\beta(s)^{-1}, \quad \bar{a}(s,t) = {}^{\beta(s)}[\Delta^{\phi(s)}_{\alpha(s)} m(t)] \cdot {}^{\beta(s)\cdot{}^{\phi(s)}\beta(t)}[a(s,t)]\,.$$

The last formula corresponds to an action

$$\nabla : C^1(G,E) \times C^2(G,B) \to C^2(G,B)$$

such that the following distributivity law holds

$$(4.12) \qquad m \nabla [(\alpha,a) * \underline{\underline{k}}] = (m \nabla (\alpha,a)] * [m \nabla \underline{\underline{k}}]\,.$$

Similarly, so as to clarify the transformation of the orbit Ω into another one $\Omega' = {}^r\Omega$ by the third action, we observe two things. One is that moving $\underline{\underline{k}}$ into $(\alpha,a) * k$ and then twist into ${}^\pi[(\alpha,a) * k]$ is equivalent to first twisting $\underline{\underline{k}}$ into ${}^\pi\underline{\underline{k}}$ and then move through $({}^\pi\alpha, {}^\pi a)$, so that we have another distributivity law reading

$$(4.13) \qquad {}^\pi[(\alpha,a) * \underline{\underline{k}}] = {}^\pi(\alpha,a) * {}^\pi\underline{\underline{k}}\,.$$

Another is that moving $\underline{\underline{k}}$ into $m \nabla \underline{\underline{k}}$ and then twist into ${}^\pi(m \nabla k)$ is equivalent to first twist $\underline{\underline{k}}$ into ${}^\pi\underline{\underline{k}}$ and then move into $({}^\pi m) \nabla ({}^\pi\underline{\underline{k}})$ where $({}^\pi m)(s) = {}^\pi[m(s)]$. We thus have a third distributivity law:

$$(4.14) \qquad {}^\pi(m \nabla \underline{\underline{k}}) = ({}^\pi m) \nabla ({}^\pi\underline{\underline{k}})\,.$$

The details involve long computations which we leave for a subsequent publication.

Now, an orbit ω, Ω or $\boldsymbol{\Omega}$ will be called <u>neutral</u> if it contains a neutral 3-cocycle. It should be observed that the first action transofrms a neutral cocycle in a non-neutral one in general. However neutrality is conservative with respect to the second and third actions.

If we go back the short exact sequence (E) of (2.6), we have associated to it a super-crossed group $\mathbb{A}'$ and a connecting map

$$\boldsymbol{\Delta}^3 : Z^2(G, A'') \to \mathbf{H}^3(G, \mathbb{A}')$$

inducing finer connecting maps $\boldsymbol{\Delta}^3$ and Δ^3 fitting into the diagram

$$
(4.15) \qquad
\begin{array}{ccccc}
Z^2(G,A) & \longrightarrow & Z^2(G,A'') & \xrightarrow{\boldsymbol{\Delta}^3} & \mathbf{H}^3(G,\mathbb{A}') \\
\downarrow & & \downarrow & & \downarrow \\
\mathbb{H}^2(G,A) & \longrightarrow & \mathbb{H}^2(G,A'') & \xrightarrow[\boldsymbol{\Delta}^3]{} & \mathbb{H}^3(G,\mathbb{A}') \\
\downarrow & & \downarrow & & \downarrow \\
H^2(G,A) & \longrightarrow & H^2(G,A'') & \xrightarrow[\Delta^3]{} & H^3(G,\mathbb{A}')
\end{array} .
$$

Moreover the rows in this diagram are "exact" if we endow the sets in the last column with the special subset of neutral classes.

This way, the problem which motivated the construction of a three dimensional cohomology theory is solved. But the lector will easily define super-crossed groups $\mathbb{A}$, $\mathbb{A}''$ associated with (2.6) so that, in some sense we get exact prolongations of the sequences (2.7) and (2.8) by

(4.16) $\cdots \to \mathbb{H}^2(G,A) \to \mathbb{H}^2(G,A'') \to \mathbb{H}^3(G,\mathbf{A}') \to \mathbb{H}^3(G,\mathbf{A}) \to \mathbb{H}^3(G,\mathbf{A}'')$,

(4.17) $\to H^2(G,A) \to H^2(G,A'') \to H^3(G,\mathbf{A}') \to H^3(G,\mathbf{A}) \to H^3(G,\mathbf{A}'')$,

together with an exact sequence

$$* \to Z^2(G,A') \to Z^2(G,A) \to Z^2(G,A'') \to \mathbf{H}^3(G,\mathbf{A}') \to \mathbf{H}^3(G,\mathbf{A}) \to \mathbf{H}^3(G,\mathbf{A}'') .$$

He will also be able to define the general concept of a short exact sequence of super-crossed group

$$\mathbf{A}' \xrightarrow{\text{I}} \mathbf{A} \xrightarrow{\text{J}} \mathbf{A}''$$

and construct the associated exact sequences generalizing the exact sequences (4.16) and (4.17). And then new problems (b) arise, the solution to which will require more research.

To that effect it should be reminded that for a crossed group $A = (H,\rho,\Pi,\phi)$ it is useful to consider the group $\Omega = \Pi/\rho H$. Then it turns out that $\mathbb{H}^2(G,A)$ has a natural map

$$\Theta : \mathbb{H}^2(G,A) \to \underline{\text{Hom}}(G,\Omega)$$

the fibers of which are acted upon freely by some abelain groups $H^2(G, Z_H)$ of the classical theory where Z_H is the center of A and therefore an abelian group.

Some similar structure on $\mathbb{H}^3(G,\mathbf{A})$ will undoubtedly show up. For example, if we look at formula (3.20.II), it is clear that it will be useful to consider the quotient group $\Theta = \Pi/\Lambda$ in which the pair (ϕ,η) induces obviously a 2-cocycle. Looking further at formulas (4.3), (4.5) as well as (4.9), (4.11) we obviously have canonical maps

$$\mathbb{H}^3(G, \mathbb{A}) \to \mathbb{H}^2(G, \Theta) ,$$

$$H^3(G, \mathbb{A}) \to H^2(G, \Theta) .$$

There, of course, Θ should be replaced by some crossed group derived from a finer analysis of the situation.

We have already seen that for $s, t \in G$, $K(s,t)$ is an inverse crossed homomorphism and then formula (3.15) shows that there is a canonical map

$$\mathbb{H}^3(G, \mathbb{A}) \to \underline{\mathrm{App}}(G \times G, H^1(\Lambda, B)) .$$

The latter certainly will induce maps going from $\mathbb{H}^3(G, \mathbb{A})$ and $H^3(G, \mathbb{A})$ to some quotient of $\underline{\mathrm{App}}(G \times G, H^1(\Lambda, B))$, the nature of this quotient being told by formulas (4.4) and (4.10) which should be interpreted. A rough look at (4.4) seems to show that these quotients will look as an $\mathbb{H}^2(G, ?)$ and $H^2(G, ?)$ generalizing what we already know. More precisely these sets should represent a sort of twisted 2-cohomology as already touched briefly in [3].

We hope to have shown that the subject is worth more careful consideration.

BIBLIOGRAPHY

[1] P. Dedecker - 0, Comptes rendus, 247(1958), 1160-62; I, ibid., 257(1963), 2384-87; II, ibid., 258(1964), 1117-20; III, ibid., 258(1964) 4891-93; IV, ibid., 259(1964), 2054-57; V, ibid., 260(1965), 4137-39.

[2] P. Dedecker - Cohomologie non abélienne, mimeographié, Fac. Sci. Lille, 1ère éd. 1963-64; 2ème éd. augmentée, Sept. 1965.

[3] P. Dedecker et A. Frei - VI, Comptes rendus, 262(1966), 1298-1301; VII, ibid., 263(1966), 203-206.

[4] J. Giraud - Comptes rendus, 260(1965), 2392-94 et 2666-68.

[5] J. Giraud - Cohomologie non abélienne, Thèse, Faculté des Sciences de Paris; mimeographed, Columbia University, 1966.

[6] J. P. Serre - Cohomologie galoissienne, Lecture notes no. 5, Springer Verlag.

H-SPACES

by

R. R. Douglas, P. J. Hilton, F. Sigrist

1. INTRODUCTION

Let $\mathcal{C}$ be the homotopy category of pointed, connected, finite CW-complexes. Recall that an H-space (in $\mathcal{C}$) is a pair (X,m), where $m \in \mathcal{C}(X \times X, X)$ such that

$$\begin{array}{ccc} X \times X & & \\ \uparrow i & \searrow^{m} & \\ & & X \\ & \nearrow_{\nabla} & \\ X \vee X & & \end{array}$$

commutes. (∇ is the homotopy class of the "folding" map.) In other words, m is a multiplication on X, having a two-sided homotopy unit. If in addition, the following diagram

$$\begin{array}{ccc} X \times X \times X & \xrightarrow{1 \times m} & X \times X \\ \downarrow{\scriptstyle m \times 1} & & \downarrow{\scriptstyle m} \\ X \times X & \xrightarrow{m} & X \end{array}$$

commutes, then we say that (X,m) is homotopy-associative.

In attempting to classify the H-spaces in $\mathcal{C}$, one first asks which simple objects in $\mathcal{C}$ (i.e. having few cells) support an (homotopy-associative) H-space multiplication. This question, restricted to spheres, is equivalent to the "Hopf invariant one" problem.

<u>Theorem: (Adams [1])</u>

<u>If S^n is an H-space</u>, then $n = 1, 3,$ <u>or</u> 7.

Remark. S^1 and S^3 are Lie groups, and S^7 is known to support no homotopy-associative multiplications.

Let E be (the total space of) a q-sphere bundle over the n-sphere (q, n positive). What restrictions are placed on the pair (q,n) in case E is an H-space (respectively, homotopy-associative H-space)? These two questions are answered by the following two theorems.

Theorem 1: [3,4]

If E is an H-space, then either

(i) $q, n \in \{1, 3, 7\}$, or

(ii) $(q,n) = (1,2)$, $(3,4)$ or $(3,5)$.

Remark: These restrictions are the best possible, as E may be the product bundle in case (i), and E may be S^3, S^7 or SU(3), respectively, in case (ii).

Theorem 2: [5]

If E is a homotopy-associative H-space, then either

(i) $q, n \in \{1,3\}$, or

(ii) $(q,n) = (1,2)$, $(3,5)$ or $(3,7)$.

Remark: Again, the restrictions on (q,n) are the best possible; product bundles in case (i) and SO(3), SU(3) or Sp(2), respectively, in case (ii), provide examples which are all Lie groups.

Until recently the only known H-spaces in $\mathfrak{C}$ were the compact Lie groups, S^7, RP^7, and products of these; moreover, it had been mildly conjectured that these were the only H-spaces in $\mathfrak{C}$ (up to homotopy type). A new H-space, constructed in this paper as a smooth closed 10-manifold which is a 3-sphere bundle over S^7, is a counterexample to this conjecture. It arises as a by-product of a systematic search for non-cancellation examples in products of manifolds. More precisely, two differentiable manifolds E_α and E_β are constructed, of different homotopy type, having the property that their respective products with S^3 are diffeomorphic. The new H-space thus emerges as a direct factor in a Lie group. The content of Section 3 is joint work with J. Roitberg.

2. PROOFS OF THEOREMS 1 AND 2

Suppose $S^q \longrightarrow E \longrightarrow S^n$ is a fibration and E is an H-space. If $q \geqq n$, it follows that there is a cross-section $\sigma : S^n \longrightarrow E$, and therefore S^n is an H-space. Moreover, Whitehead products on S^q are sent to Whitehead products on E by a monomorphism. Thus, S^q has trivial Whitehead products, and it follows that S^q is an H-space.

Without loss of generality, we now assume $q < n$. Using the usual spectral sequence, we compute $H^*(E;Z_2)$. If $q = n - 1$, $H^*(E;Z_2)$, may be an exterior algebra on one generator of dimension $2n - 1$. If $H^*(E;Z_2) \approx H^*(S^{2n-1};Z_2)$, then it follows from [1] that $2n - 1 = 3$ or 7 (recall $n > 1$); thus, $(q,n) = (1,2)$ or $(3,4)$. If $H^*(E;Z_2)$ is not an exterior algebra on one generator, then $H^*(E;Z_2) \approx H^*(S^q \times S^n;Z_2)$, and, by [2], it suffices to show that (q,n) can not be $(7,11)$ or $(7,15)$.

If $S^7 \longrightarrow E \longrightarrow S^n$ is a fibration, with $n = 11$ or 15, let P_2E be the projective plane of the H-space E; then $KU(P_2E)$ may be thought of as a (torsion-free) filtered, truncated polynomial algebra on two generators α, β (of filtrations 8 and $n + 1$, respectively), truncated at height three. $\psi^2\psi^3(\alpha) = \psi^3\psi^2(\alpha)$ and $\psi^2\psi^3(\beta) = \psi^3\psi^2(\beta)$, together with $\psi^2(\alpha) - \alpha^2 \equiv 0 \pmod 2$ and $\psi^2(\beta) - \beta^2 \equiv 0 \pmod 2$, lead to a contradiction (for details of proof see [4]).

Now suppose $S^q \xrightarrow{i} E \xrightarrow{p} S^n$ is a fibration and E is a homotopy-associative H-space. Theorem 2 will follow from Theorem 1, if we can show that (q,n) can not be $(3,4)$, $(1,7)$, $(7,1)$, $(7,3)$, or $(7,7)$.

Lemma 1

If $S^3 \longrightarrow X \longrightarrow S^4$ is a fibration and X is an H-space, then X has the homotopy type of S^7.

Lemma 2

If $S^q \longrightarrow X \longrightarrow S^n$ is a fibration, with $(q,n) = (1,7)$, or $(7,1)$, then the universal covering $\widetilde{X}$ of X has the homotopy type of S^7.

Remark: The universal covering of a homotopy-associative H-space is again a homotopy-associative H-space.

Lemma 3

If X is a homotopy-associative H-space and $H^*(X;Z) \approx H^*(S^3 \times S^7;Z)$, then the Steenrod operation $P_3^1 : H^3(X;Z_3) \longrightarrow H^7(X;Z_3)$ is non-trivial.

Now let $(q,n) = (7,3)$ and $X = E$. The Serre exact sequence for cohomology implies that

$$i^* : H^7(E;Z_3) \longrightarrow H^7(S^7;Z_3)$$

is an isomorphism, which contradicts the conclusion of Lemma 3.

Finally, $(q,n) \neq (7,7)$ for essentially the same reason that S^7 can not be a homotopy-associative H-space (i.e., $P_3^4 = P_3^1 \cdot P_3^3$ in the mod 3 Steenrod algebra).

3. S^3-BUNDLES

We consider principal S^3-bundles over S^n. Such bundles are classified by elements $\alpha \in \pi_{n-1}(S^3)$. Let B be the classifying space and let $\alpha \longleftrightarrow \alpha_0$ under the canonical isomorphism $\pi_{n-1}(S^3) \cong \pi_n(B)$. We adopt the notation

$$\begin{array}{ccc} S^3 \subseteq E_\alpha & & \\ \downarrow P_\alpha & & \\ S^n & \xrightarrow{\alpha_0} & B, \end{array}$$

deliberately confusing maps and homotopy classes.

<u>Proposition 1 (James-Whitehead)</u>

<u>E_α has a CW-structure</u>

$$E_\alpha = (S^3 \cup_\alpha e^n) \cup e^{n+3} \quad .$$

<u>Theorem 3</u>

$$E_\alpha \simeq E_\beta \Longleftrightarrow \beta = \pm\alpha \quad .$$

This follows easily from Proposition 1.

Now consider the fibre-product diagram

$$\begin{array}{ccccc} E_{\alpha\beta} & \longrightarrow & E_\beta & & \\ \downarrow & & \downarrow P_\beta & & \\ E_\alpha & \xrightarrow{P_\alpha} & S^n & \xrightarrow{\beta_0} & B \quad . \end{array}$$

<u>Proposition 2</u>

$E_{\alpha\beta} = E_{\beta\alpha}$; <u>and</u> $E_{\alpha\beta} = E_\alpha \times S^3$ <u>if</u> $\beta_0 \cdot P_\alpha = 0$.

<u>Theorem 4</u>

<u>Let</u> $\beta = \ell\alpha$. <u>Then</u> $\beta_0 \cdot P_\alpha = 0$ <u>if</u> $\frac{\ell(\ell-1)}{2}\omega \cdot \Sigma^3\alpha = 0$, <u>where</u> $\omega \in \pi_6(S^3)$ <u>measures the non-commutativity of</u> S^3.

To prove this we consider the Puppe sequence of the inclusion of the fibre in E_α,

$$S^3 \xrightarrow{i} E_\alpha \longrightarrow S^n + S^{n+3} \xrightarrow{u} S^4 \quad .$$

We readily identify u, in terms of its components, as

$$u = \langle \Sigma\alpha, \gamma \cdot \Sigma^4\alpha \rangle,$$

where $\gamma \in \pi_7(S^4)$ is the Hopf map. Now $P_\alpha = qj$,

$q: S^n + S^{n+3} \longrightarrow S^n$ and

$$\alpha_0 \cdot q = e \cdot u, \quad \text{or} \quad \alpha_0 \cdot q = \langle \alpha_0, 0 \rangle,$$

where $e \in \pi_4(B)$ is adjoint to $1 \in \pi_3(S^3)$. Then

$$\ell\alpha_0 \cdot q = \langle \ell\alpha_0, 0 \rangle$$

$$\text{and} \quad \ell e \cdot u = \langle \ell\alpha_0, \ell e \cdot \gamma \cdot \Sigma^4\alpha \rangle$$

$$= \langle \ell\alpha_0, \frac{\ell(\ell-1)}{2} [e,e] \cdot \Sigma^4\alpha \rangle \quad .$$

Now the Whitehead product [e,e] is adjoint to ω. Thus,

$$\ell\alpha_0 \cdot q = \ell e \cdot u \text{ , and hence } \ell\alpha_0 \cdot P_\alpha = 0,$$

if $\frac{\ell(\ell-1)}{2} \omega \cdot \Sigma^3\alpha = 0$.

Corollary 1. Let α be of order k, $k_0 = \gcd(k,24)$, ℓ prime to k, $\ell \equiv 1 \bmod k_0$, $\beta = \ell\alpha$. Then

$$E_\alpha \times S^3 = E_\beta \times S^3 \quad .$$

Thus, for example, if $k = p$, a prime $\neq 2,3$, and if $\beta = \ell\alpha$ where ℓ is prime to p and $\ell \not\equiv \pm 1 \bmod p$, then $E_\alpha \times S^3 = E_\beta \times S^3$, although $E_\alpha \not\simeq E_\beta$.

Corollary 2. Let $\alpha = \omega$, $\beta = 7\omega$. Then

$$E_\omega \not\simeq E_{7\omega} \;, \quad E_\omega \times S^3 = E_{7\omega} \times S^3 \quad .$$

This follows from Theorems 3 and 4 and the fact that $\pi_9(S^3) = Z_3$. Now $E_\omega = Sp(2)$, so $E_{7\omega}$ is an H-space. On dimensionality grounds it follows that it cannot be any of the known H-manifolds; it is the total space of a principal S^3-bundle over S^7.

REFERENCES

[1] Adams, J. F., "On the non-existence of elements of Hopf invariant one", *Ann. of Math.* 72; 20-104. (1960).

[2] Adams, J. F., "H-spaces with few cells", *Topology* 1; 67-72. (1962).

[3] Douglas, R. R. and F. Sigrist, "Sphere bundles over spheres which are H-spaces", *Rendic. Acc. Naz. Lincei* 44; No.4. (1968).

[4] Douglas, R. R. and F. Sigrist, "Sphere bundles over spheres and H-spaces", (To appear).

[5] Douglas, R. R. and F. Sigrist, "Homotopy-associative H-spaces which are sphere bundles over spheres", (To appear).

CONSTRUCTION DE STRUCTURES LIBRES

Charles Ehresmann

INTRODUCTION.

Le but de cet article est de donner un critère d'existence de p-structures libres et des applications de ce critère, qui fait intervenir un foncteur auxiliaire P dont p est une restriction; par exemple si p est le foncteur d'oubli relatif à l'univers M_o de la catégorie des homomorphismes entre structures d'un certain type, P pourra être le foncteur analogue relatif à un univers $\bar{M}_o$ auquel appartient M_o. Contrairement au théorème d'existence d'adjoint de Freyd, ce critère impose des propriétés (telles que l'existence d'un "assez grand nombre" de produits) sur P, non sur le foncteur donné p. Les hypothèses peuvent en être généralisées (voir [1]), mais ici nous avons cherché à indiquer des conditions simples, réalisées dans la plupart des exemples.

Comme application, nous obtenons des théorèmes d'existence de limites inductives ou de structures quasi-quotients (voir aussi [1]), et un théorème de complétion "maximale" d'une catégorie en une catégorie à limites projectives et inductives d'une certaine espèce, avec conservation de limites données. Nous avons montré ailleurs [2] que ce dernier résultat a des conséquences intéressantes pour l'étude d'une notion générale de "structure algébrique" définie comme réalisation, dans une catégorie quelconque, d'une esquisse (i.e. d'un graphe multiplicatif muni d'une famille de transformations naturelles).

0. *QUELQUES RAPPELS.*

Nous nous plaçons dans le cadre de la Théorie des ensembles avec existence d'au moins deux univers M_o et $\hat{M}_o$ tels que $M_o \in \hat{M}_o$. Mais en fait la plupart des raisonnements seraient aussi valables dans une Théorie avec ensembles et classes, en prenant pour M_o la classe de tous les ensembles et pour éléments de $\hat{M}_o$ des classes. Dans le dernier §, l'axiome du choix est librement utilisé dans M_o.

La terminologie et les notations sont ceux de [3], dans l'index duquel se trouvent les mots non explicitement définis ici. Les autres notions que nous allons rappeler figurent dans le cours [4].

Univers. Ensemble M_o d'ensembles tel que:

1° Si E appartient à M_o, l'ensemble $P(E)$ de ses parties est un élément et une partie de M_o.

2° Si $(E_i)_{i \in I}$ est une famille d'éléments de M_o indexée par un élément I de M_o, sa réunion appartient à M_o.

3° Il existe un ensemble infini appartenant à M_o. (Contrairement à la définition de Grothendieck, nous n'exigeons pas que M_o soit un ensemble transitif).

Application. Une application f de M dans M' est désignée par le triplet $(M', \underline{f}, M)$, où $\underline{f}$ est la surjection

$$x \to f(x) \quad \text{de} \quad M \quad \text{sur} \quad f(M) \subset M'.$$

Si f est l'injection canonique de $M \subset M'$ dans M', on écrit $f = (M', \iota, M)$.

$\mathcal{M}$ désigne toujours la catégorie pleine des applications associée à un univers $\mathcal{M}_o$.

Catégorie. La catégorie (C,κ), où κ est la loi de composition $(y,x) \to y.x$ de $M \subset C\times C$ dans C, est notée $C^{\bullet}$ (ou, parfois, C), et l'on pose $M = C^{\bullet} * C^{\bullet}$. La classe des unités de $C^{\bullet}$ est $C^{\bullet}_o$, les applications source et but α et β , le graphe (C,β,α) sous-jacent $[C^{\bullet}]$, le groupoïde des éléments inversibles $C^{\bullet}_{\gamma}$. Si A et B sont des parties de C, l'ensemble des composés $y.x$ tels que $y \in B$ et $x \in A$ est désigné par $B.A$; on pose:

$$\{b\}.A = b.A \quad \text{et} \quad B.\{a\} = B.a \, .$$

Foncteur. La surjection définissant le foncteur F de $C^{\bullet}$ vers $H^{\bullet}$ est notée $\underline{F}$, de sorte que $F = (H^{\bullet},\underline{F},C^{\bullet})$, et l'on écrit: $\alpha(F) = C^{\bullet}$, $\beta(F) = H^{\bullet}$.

Transformation naturelle. La catégorie (longitudinale) des transformations naturelles entre foncteurs de $C^{\bullet}$ vers $H^{\bullet}$ est notée $\mathbf{n}(H^{\bullet},C^{\bullet})^{\text{⊡}}$. Soit T un de ses éléments , (F',t,F). Si p est un foncteur de $H^{\bullet}$ vers $K^{\bullet}$, la transformation naturelle $(p.F',\underline{p}t,p.F)$, est notée pT. Si F' est le foncteur $\hat{e}$ constant sur l'unité e de $H^{\bullet}$ et si $h' \in H.e$, soit $h'T$ la transformation naturelle

$$(\widehat{\beta(h')},t',F), \quad \text{où } t'(i) = h'.t(i) \text{ pour tout } i \in C^{\bullet}_o.$$

De même, si $F = \hat{e}$ et si $h \in e.H$, on dénote Th la transformation naturelle

$$(F',t'',\widehat{\alpha(h)}), \quad \text{où } t''(i) = t(i).h \text{ pour tout } i \in C^{\bullet}_o.$$

Produits. Si $(e_i)_{i\in I}$ est une famille d'unités de $H^{\bullet}$ admettant e pour produit dans $H^{\bullet}$, la projection canonique de e vers e_i étant p_i, on appelle $((p_i)_{i\varepsilon I},e)$ un produit naturalisé dans $H^\bullet$; dans ce cas, pour toute famille $(f_i)_{i\varepsilon I}$ de morphismes f_i de même source et de but e_i, l'unique f tel que

$$f.p_i = f_i \quad \text{pour tout} \quad i \varepsilon I$$

est noté $[f_i]_{i\varepsilon I}$.

Soit L un ensemble d'ensembles; $H^\bullet$ est à L-produits si toute famille d'unités $(e_i)_{i\varepsilon I}$, où $I \varepsilon L$, admet un produit dans $H^\bullet$. Un foncteur p de $H^\bullet$ vers $K^\bullet$ est à L-produits si $H^\bullet$ et $K^\bullet$ sont à L-produits et si

$$((p(p_i))_{i\varepsilon I}, p(e))$$

est un produit naturalisé dans $K^\bullet$ lorsque $((p_i)_{i\varepsilon I},e)$ en est un dans $H^\bullet$.

Noyaux. Une catégorie $H^\bullet$ est à noyaux si tout couple (h',h), où $h \varepsilon H$ et $h' \varepsilon \beta(h).H.\alpha(h)$, admet un noyau dans $H^\bullet$. Un foncteur p de $H^\bullet$ vers $K^\bullet$ est à noyaux si $H^\bullet$ et $K^\bullet$ sont à noyaux et si $p(j)$ est un noyau de $(p(h),p(h'))$ dans $K^\bullet$ lorsque j est un noyau de (h,h') dans $H^\bullet$.

Limites [1]. Une transformation naturelle $T = (\hat{e},t,F)$ telle que e soit une limite inductive de F (notée Lim F) et que $t(i)$ soit, pour tout $i \varepsilon \alpha(F)_o$, l'injection canonique de $F(i)$ vers e est appelée limite inductive naturalisée (de F). Dans ce cas, pour toute transformation naturelle $T' = (\hat{e}',t',F)$, on note $\lim^T T'$ (ou simplement $\lim T'$) l'unique k tel que $kT = T'$.

Définitions duales: limite projective naturalisée T et $\varprojlim^T T'$.

Soit $\mathcal{I}$ un ensemble de catégories. Une application associant à certains (resp. à tous les) foncteurs F de $C^\bullet \varepsilon \mathcal{I}$ vers $H^\bullet$ une limite inductive naturalisée de F est dite application I-limite inductive naturalisée partielle (resp. naturalisée) sur $H^{\bullet}$. Même définition pour une application I-limite projective naturalisée (partielle) sur $H^{\bullet}$.

p-*structure libre*. Soit p un foncteur de $H^{\bullet}$ vers $K^{\bullet}$. On appelle p-projecteur un couple $(s,g) \in H_o^{\bullet} \times K$ tel que $p(s) = \beta(g)$ et que, si $(s',g') \in H_o^{\bullet} \times K$ et si g' appartient à $p(s').K.\alpha(g)$, il existe un et un seul $h \in s'.H.s$ vérifiant $p(h).g = g'$; on dit alors que s est une p-structure libre associée à $\alpha(g)$.

Si, à toute unité e de $K^{\bullet}$, est associée une p-structure libre $p'(e)$, l'application $e \rightarrow p'(e)$ se prolonge en un foncteur adjoint p' de p.

Si p est le foncteur injection canonique vers $K^{\bullet}$ de sa sous-catégorie $H^{\bullet}$ et si p admet un adjoint, $K^{\bullet}$ est dite catégorie à H-projections.

p-*injection*. Soit p un foncteur de $H^{\bullet}$ vers $K^{\bullet}$. Un élément j de H est une p-injection si, pour tout élément h de $\beta(j).H$ tel que $p(h) = p(j).k$, où $k \in K$, il existe un et un seul $h' \in H$ vérifiant:

$$h = j.h' \quad \text{et} \quad p(h') = k.$$

Si de plus $p(j)$ est un monomorphisme de $K^{\bullet}$, on appelle j un p-monomorphisme.

Par dualité, on obtient les p-surjections et les p-épimorphismes.

Structures quasi-quotients. Soit p un foncteur de $H^{\bullet}$ vers la catégorie M d'applications. Soit s une unité de $H^{\bullet}$ (ou p-structure sur l'ensemble $p(s)$), et r une relation d'équivalence sur $p(s)$. On se donne une sous-catégorie pleine $H'^{\bullet}$ de $H^{\bullet}$. On dit que $\hat{s} \in H'$ est une (H',p)-structure quasi-quotient de s par r (voir [5]) s'il existe un élément j de $\hat{s}.H.s$ (appelé (H',p)-quasi-surjection) vérifiant les conditions:

1° $p(j)$ est compatible avec r.

2° Si $h \in H'.H.s$ et si $p(h)$ est compatible avec r, il existe un et un seul $h' \in H'$ tel que $h = h'.j$.

Cette définition signifie que $\hat{s}$ est une H'-projection de (s,r) dans une certaine catégorie p^{r}.(Voir [5]).

Si, pour toute unité s de $H^{\bullet}$ et toute relation d'équivalence r sur $p(s)$, il existe une p-structure quasi-quotient (i.e. une (H,p)-structure quasi-quotient) de s par r, on dit que p est un foncteur à structures quasi-quotients.

Si j est une p-quasi-surjection définissant $\hat{s}$ comme p-structure quasi-quotient de s par r et si $p(j)$ est la surjection canonique $\tilde{r}$ de $p(s)$ sur l'ensemble quotient $p(s)/r$, on appelle $\hat{s}$ une p-structure quotient de s par r.

Cas particulier: Catégories quotients d'une catégorie.

Sous-morphismes engendrés. Soit p un foncteur de $H^{\bullet}$ vers $K^{\bullet}$, et soit X une partie de H. Si S est une unité de $H^{\bullet}$ et si $f \in p(S).K$, on dit que j est un (X,p)-sous-morphisme de S engendré par f s'il vérifie les conditions suivantes:

1° $j \in S.X$ et il existe un $k \in K$ tel que $f = p(j).k$.

2° Si $j' \in S.X$ et s'il existe un $k' \in K$ tel que $f = p(j').k'$, il existe un et un seul $h \in H$ vérifiant:

$$j'.h = j \quad \text{et} \quad p(h).k = k'.$$

Graphe multiplicatif. C'est un système multiplicatif $C^{\bullet}$ pour lequel il existe un graphe (C,β,α) ayant les propriétés:

1° Si $y.x$ est défini, on a

$$\alpha(y) = \beta(x) , \quad \alpha(y.x) = \alpha(x) \quad \text{et} \quad \beta(y.x) = \beta(y).$$

2° Pour tout $x \in C$, les composés $x.\alpha(x)$ et $\beta(x).x$ sont définis et égaux à x.

La classe $\alpha(C)$ des unités de $C^{\bullet}$ est notée $C^{\bullet}_{o}$.

Néofoncteur. C' est un triplet $F = (K^{\bullet}, \underline{F}, C^{\bullet})$, où $C^{\bullet}$ et $K^{\bullet}$ sont des graphes multiplicatifs et où $(K, \underline{F}, C)$ est une application f telle que:

1° $f(C^{\bullet}_{o}) \subset K^{\bullet}_{o}$.

2° Si $y.x$ est défini dans $C^{\bullet}$, le composé $f(y).f(x)$ est défini dans $K^{\bullet}$ et l'on a:

$$f(y.x) = f(y).f(x).$$

Nous désignons par $\mathcal{N}'$ (resp. par $\mathcal{F}$) la catégorie des néo-foncteurs (resp. des foncteurs) associée à l'univers $\mathcal{M}_{o}$.

1. THEOREME D'EXISTENCE DE STRUCTURES LIBRES.

Soient p un foncteur de $H^{\bullet}$ vers $K^{\bullet}$ et e une unité de $K^{\bullet}$. Nous allons donner des conditions suffisantes pour qu'il existe une p-structure libre associée à e.

HYPOTHESES. 1° p est la restriction d'un foncteur P de $\hat{H}^{\bullet}$ vers $\hat{K}^{\bullet}$ à des sous-catégories pleines $H^{\bullet}$ de $\hat{H}^{\bullet}$ et $K^{\bullet}$ de $\hat{K}^{\bullet}$.

2° Désignons par I l'ensemble des couples (s,g), où $s \in H^{\bullet}_{o}$ et $g \in p(s).K.e$; le couple $i = (s,g)$ sera noté (s_i, g_i). On suppose qu'il existe un couple $(S, \hat{g})$, où S est une unité de $\hat{H}^{\bullet}$ et $\hat{g} \in P(S).\hat{K}.e$, tel que, pour tout $i \in I$, il existe un $p_i \in s_i.\hat{H}.S$ vérifiant $P(p_i).\hat{g} = g_i$.

3° Soit X un ensemble de monomorphismes de $\hat{H}^{\bullet}$ tel que les éléments de $P(X)$ soient des monomorphismes de $\hat{K}^{\bullet}$. Il existe un $(X.H^{\bullet}_{o}, P)$-sous-morphisme j de S engendré par $\hat{g}$; on note $\hat{s} = \alpha(\hat{j})$.

4° Si h et h' sont deux éléments de H de source $\hat{s}$ et de même but, il existe un monomorphisme n de $H^{\bullet}$ tel que $h.n = h'.n$, $X.n \subset X$ et que $p(n)$ soit un noyau de $(p(h'), p(h))$ dans $K^{\bullet}$.

PROPOSITION 1. Si les hypothèses 1, 2, 3 et 4 sont vérifiées, $\hat{s}$ *est une p-structure libre associée à* e.

Δ. D'après la définition d'un $(X.H^{\bullet}_o, P)$-sous-morphisme engendré, il existe un $k \in \hat{K}$ tel que $\hat{g} = P(j).k$; puisque $\alpha(k) = \alpha(\hat{g}) = e$ et $\beta(k) = P(\hat{s}) \in K$ et que $K^{\bullet}$ est une sous-catégorie pleine de $\hat{K}^{\bullet}$, on a $k \in K$. Montrons que $(\hat{s}, k)$ est un p-projecteur. En effet, si $i = (s_i, g_i)$ est un élément de I, on trouve, avec les notations de la condition 2,

$$g'_i = p_i.j \in s_i.\hat{H}.\hat{s} = s_i.H.\hat{s}$$

et

$$p(g'_i).k = P(p_i).P(j).k = P(p_i).\hat{g} = g_i \;.$$

D'autre part, supposons qu'il existe aussi $g''_i \in s_i.H.\hat{s}$ avec $p(g''_i).k = g_i$. En utilisant la condition 4 pour $h = g'_i$ et $h' = g''_i$, on obtient un monomorphisme n tel que $g'_i.n = g''_i.n$, $X.n \subset X$ et que $p(n)$ soit un noyau de $(p(g'_i), p(g''_i))$ dans $K^{\bullet}$. Cette dernière condition entraîne l'existence d'un k' vérifiant $p(n).k' = k$, car

$$k \in K \text{ et } p(g'_i).k = g_i = p(g''_i).k \;.$$

Il s'ensuit $j.n \in X$ et

$$P(j.n).k' = P(j).P(n).k' = P(j).k = \hat{g}.$$

Par définition de j, il existe un $n' \in H$ tel que $j = (j.n).n'$. Or, j étant un monomorphisme, il résulte de cette égalité $n.n' = \hat{s}$, d'où $n' = n^{-1}$ vu que n est un

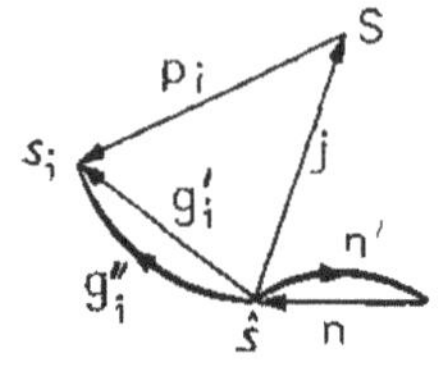

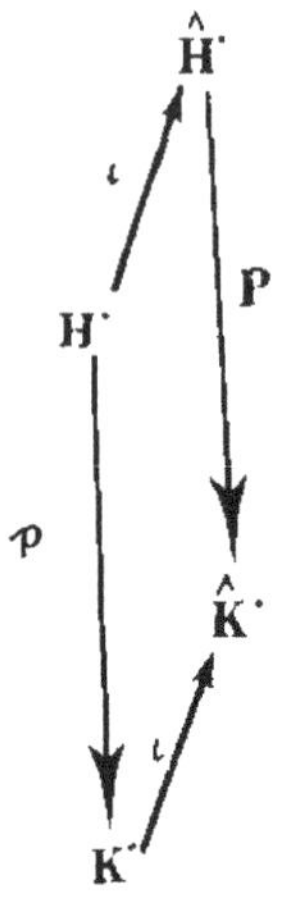

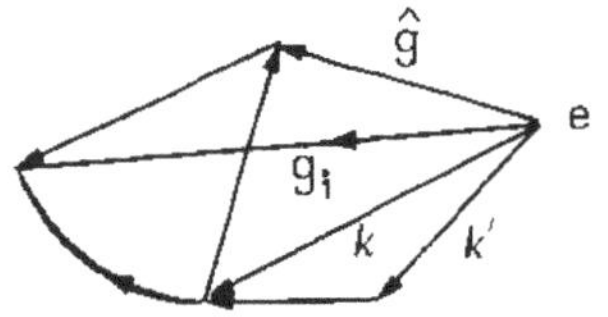

monomorphisme. Donc n est inversible, et $g'_i = g''_i$. Ceci prouve que $(\hat{s},k)$ est un p-projecteur. ∇

CAS PARTICULIERS. A° Supposons que la condition 1 soit vérifiée et qu'il existe un produit naturalisé $((p_i)_{i\in I},S)$ de $(s_i)_{i\in I}$ dans P (i.e. un produit naturalisé dans $\hat{H}^{\bullet}$ tel que $((P(p_i))_{i\in I},P(S))$ soit un produit naturalisé dans $\hat{K}^{\bullet}$); en prenant pour $\hat{g}$ l'unique élément $\hat{g} = [g_i]_{i\in I}$ tel que $P(p_i).\hat{g} = \hat{g}_i$ pour tout $i \in I$, la condition 2 est remplie.

B° La condition 4 est satisfaite lorsque p est un foncteur à noyaux et que $X.n \subset X$ pour tout noyau n.

C° Supposons vérifiées les hypothèses 1 et 2, et soit Q un foncteur de $\hat{K}^{\bullet}$ vers $L^{\bullet}$. Soit X un ensemble de monomorphismes de $\hat{H}^{\bullet}$ tel que P(X) soit formé de monomorphismes de $\hat{K}^{\bullet}$. S'il existe un $(X.H^{\bullet}_o,Q.P)$-sous-morphisme j de S engendré par $Q(\hat{g})$ et si P(j) est une Q-injection, alors j est un $(X.H^{\bullet}_o,P)$-sous-morphisme de S engendré par $\hat{g}$, de sorte que l'hypothèse 3 est remplie. En effet, il existe un k tel que $Q.P(j).k = Q(\hat{g})$ et, $P(j)$ étant un Q-monomorphisme, il existe un unique $\hat{g}'$ vérifiant

$$Q(\hat{g}') = k \quad \text{et} \quad P(j).\hat{g}' = \hat{g}.$$

Par ailleurs, si $j' \in S.X$ et si $P(j').\hat{g}'' = \hat{g}$, on trouve $Q.P(j).Q(\hat{g}'') = Q(\hat{g})$ et, par définition d'un $(X.H^{\bullet}_{o},Q.P)$-sous-morphisme, il existe un $h \in H$ vérifiant $j'.h = j$; il en résulte $P(h).\hat{g}' = \hat{g}''$, ce qui prouve l'affirmation.

2. *EXISTENCE DE STRUCTURES QUASI-QUOTIENTS.*

Nous nous donnons deux univers M_o et $\hat{M}_o$ tels que $M_o \in \hat{M}_o$ et $M_o \subset \hat{M}_o$. Soient M et $\hat{M}$ les catégories d'applications correspondantes. Nous supposons que Q est un foncteur de $\hat{H}^{\bullet}$ vers $\hat{M}$ et que q est un foncteur de $H^{\bullet}$ vers M restriction de Q à la sous-catégorie pleine $H^{\bullet}$ de $\hat{H}^{\bullet}$. Soit $H'^{\bullet}$ une sous-catégorie pleine de $H^{\bullet}$. Nous allons indiquer des conditions suffisantes pour l'existence d'une (H',q)-structure quasi-quotient de s par r, où s est une unité de $H^{\bullet}$ et r une relation d'équivalence sur $q(s)$.

HYPOTHESES. 1° Soit I l'ensemble des $i \in H'^{\bullet}_{o}.H.s$ tels que $q(i)$ soit compatible avec r. Il existe un produit naturalisé $((p_i)_{i\in I},S)$ de $(\beta(i))_{i\in I}$ dans Q. Soit $\hat{g} = [i]_{i\in I}$ l'unique élément de $\hat{H}$ tel que $p_i.\hat{g} = i$ pour tout $i\in I$.

2° X étant un ensemble donné de Q-monomorphismes, il existe un $(X.H'^{\bullet}_{o},Q)$-sous-morphisme de S engendré par $Q(\hat{g})$; posons $\hat{s} = \alpha(j)$.

3° Si $h' \in H'$ et $h'' \in H'$ ont même source $\hat{s}$ et même but, il existe un noyau n de (h',h'') dans q (i.e. un noyau dans $H^{\cdot}$ tel que $q(n)$ soit un noyau de $(q(h'),q(h''))$ dans M) vérifiant $\alpha(n) \in H'$ et $X.n \subset X$.

PROPOSITION 2. Si les hypothèses précédentes sont vérifiées, $\hat{s}$ est une (H',q)-structure quasi-quotient de s par r.

Δ. Par définition de j, on a $\alpha(j) \in H'$ et il existe un k tel que $Q(j).k = Q(\hat{g})$. Comme j est un Q-monomorphisme, il existe un et un seul $g \in H$ vérifiant $Q(g) = k$ et $j.g = \hat{g}$. Si $i \in I$, dire que $Q(i)$ est compatible avec r signifie que $Q(i) = i'.\tilde{r}$, où $\tilde{r}$ est la surjection canonique de $Q(s)$ sur $Q(s)/r$; l'égalité

$$Q(\hat{g}) = [Q(i)]_{i \in I} = [i']_{i \in I}.\tilde{r}$$

entraîne que $Q(\hat{g})$ est compatible avec r; l'application $Q(j)$ étant une injection, $Q(g)$ est aussi compatible avec r. De plus, pour tout $i \in I$, on a $p_i.j \in H'$ et

$$(p_i.j).g = p_i.\hat{g} = i.$$

Par ailleurs, supposons $h'.g = h''.g$, où h' et h'' appartiennent à H'. En utilisant l'hypothèse 3, on trouve $j.n \in X$ et, n étant un noyau de (h',h''), il existe un g' tel que $n.g' = g$. Il s'ensuit

$$Q(j.n).Q(g') = Q(j).Q(g) = Q(\hat{g}),$$

de sorte que, j étant un $(X.H'^{\cdot}_{o},Q)$-sous-morphisme de S engendré par $\hat{g}$, il existe un n' vérifiant $(j.n).n' = j$. On en déduit $n.n' = \hat{s}$ et $n' = n^{-1}$, car j et n sont des monomorphismes. Donc $h' = h''$ et g est une (H',q)-quasi-surjection définissant $\hat{s}$ comme (H',q)-structure quasi-quotient de s par r. ∇

REMARQUE. La proposition 2 peut se déduire de la proposition 1 appliquée au foncteur canonique p de $H'^{\cdot}$ vers la catégorie $q^{\mathcal{h}}$ servant à définir les q-structures quasi-quotients. En effet, on montre facilement que, en identifiant $\hat{H}^{\cdot}$ à une sous-catégorie de $Q^{\mathcal{h}}$, l'élément S est un produit de $(\beta(i))_{i\epsilon I}$ dans $Q^{\mathcal{h}}$, que n est un noyau de (h',h") dans $q^{\mathcal{h}}$ et que j est un $(X.H'^{\cdot}_{o},P)$-sous-morphisme de S engendré par $(\hat{g},r)$, si P désigne le foncteur canonique de $\hat{H}^{\cdot}$ vers $Q^{\mathcal{h}}$.

Considérons toujours les foncteurs Q et q (le plus souvent nous aurons $H = \bar{Q}^{1}(\mathcal{M}))$. Soit X un ensemble de Q-monomorphismes. Nous notons $\tilde{\mathcal{M}}_o$ la saturante de $\mathcal{M}_o$ dans $\hat{\mathcal{M}}_o$, c'est-à-dire l'ensemble des éléments de $\hat{\mathcal{M}}_o$ qui sont équipotents à un élément de $\mathcal{M}_o$; on montre que $\tilde{\mathcal{M}}_o$ est un univers.

Définition. On dit que Q est $(\mathcal{M},X)$-*engendrant* si, pour tout $S \in \hat{H}^{\cdot}_o$ et tout $f \in Q(S).\tilde{\mathcal{M}}.\mathcal{M}_o$, il existe un (X,Q)-sous-morphisme de S engendré par f.

PROPOSITION 3. Q est $(\mathcal{M},X)$-engendrant si, et seulement si, pour tout $S \in \hat{H}^{\cdot}_o$ et toute partie M de Q(S) équipotente à un élément de $\mathcal{M}_o$, il existe un (X,Q)-sous-morphisme de S engendré par l'injection canonique $(Q(S),\iota,M)$.

Δ. Supposons $S \in \hat{H}^{\cdot}_o$ et $f \in Q(S).\tilde{\mathcal{M}}$. -Si Q est $(\mathcal{M},X)$-engendrant et si $f = (Q(S),\iota,M)$, où $M \in \tilde{\mathcal{M}}_o$, il existe une bijection g d'un $M' \in \mathcal{M}_o$ sur M et un (X,Q)-sous-morphisme j de S engendré par f.g. Il est clair que j est un (X,Q)-sous-morphisme de S engendré par f. - Inversement supposons vérifiée la condition de la proposition et soit $f = (Q(S),\underline{f},A) \in \hat{\mathcal{M}}.\mathcal{M}_o$. En posant $M = f(A)$, on a $M \in \tilde{\mathcal{M}}_o$, car M est équipotent au quotient de A par la relation d'équivalence associée à f, lequel quotient appartient à l'univers $\tilde{\mathcal{M}}_o$. De plus $f = i.f'$, où

$$i = (Q(S),\iota,M) \quad \text{et} \quad f' = (M,\underline{f},A).$$

Comme il existe un (X,Q)-sous-morphisme j' de S engendré par i, on obtient $i = Q(j').k$. Il s'ensuit

$$f = i.f' = Q(j').(k.f') .$$

Si $j'' \in S.X.s$ et $f = Q(j'').k'$, l'ensemble M est contenu dans $Q(j'')(Q(s))$ et $Q(j'')$ est une injection $(M',\underline{j}'',Q(s))$. Il en résulte $i = Q(j'').k''$, où $\underline{k}''$ est la restriction à M de la bijection $\underline{j}''^{-1}$. Donc il existe un h vérifiant $j' = j''.h$, et j' est un (X,Q)-sous-morphisme de S engendré par f. Ceci prouve que Q est $(\mathcal{M},X)$-engendrant. ∇

Des propositions 1, 2 on déduit:

COROLLAIRE. *Supposons que Q soit à $\mathcal{H}$-produits, que q soit à noyaux et que Q soit $(\mathcal{M},X.H^{\bullet}_o)$-engendrant. Si $X.n \subset X$ pour tout noyau n dans q, alors q admet un adjoint et q est à structures quasi-quotients.*

En effet, les hypothèses des propositions 1 et 2 sont évidemment vérifiées. ∇

Application.

Dans la plupart des exemples, les foncteurs Q et q et l'ensemble X possèdent les propriétés:

(1) Q est fidèle, $Q(s.\hat{H}^{\bullet}_{\gamma}.H^{\bullet}_o) = Q(s).\hat{\mathcal{M}}_{\gamma}.\mathcal{M}_o$ pour tout $s \in \hat{H}^{\bullet}_o$ et $X.\hat{H}^{\bullet}_{\gamma} \subset X$.

Dans la fin de ce §, nous supposons cette condition remplie et nous notons Y l'ensemble des $j \in X$ tels que $Q(j)$ soit une injection canonique ayant pour source un élément de $\tilde{M}_o$. Soit $S \in \hat{H}_o^\bullet$. Si $j \in S.X.H_o^\bullet$, il existe un $\hat{g} \in S.\hat{H}^\bullet.H_o^\bullet$ tel que $Q(\hat{g})$ soit la bijection $\underline{j}$ définissant l'injection $Q(j)$, et l'on a $j = j'.\hat{g}$, où $j' = j.\hat{g}^{-1} \in Y$. Par suite $X.H_o^\bullet = Y.\hat{H}_Y^\bullet.H_o^\bullet$. Il en résulte que, si M est une partie de $Q(S)$, alors j est un $(X.H_o^\bullet,Q)$-sous-morphisme de S engendré par $(Q(S),\iota,M)$ si, et seulement si, j' est un (Y,Q)-sous-morphisme de S engendré par $(Q(S),\iota,M)$. Ceci permet de poser la

Définition. Soient $S \in \hat{H}_o^\bullet$ et M une partie de $Q(S)$. S'il existe un (Y,Q)-sous-morphisme j de S engendré par $(Q(S),\iota,M)$, on appelle $\alpha(j)$ une $(X.H_o^\bullet,Q)$-*sous-structure de S engendrée par M.*

Grâce à la propriété (1), on déduit facilement de la proposition 3 et de la remarque précédente que Q est $(M,X.H_o^\bullet)$-engendrant si, et seulement si, pour tout $S \in \hat{H}_o^\bullet$ et toute partie M de $Q(S)$ appartenant à $\tilde{M}_o$, il existe une $(X.H_o^\bullet,Q)$-sous-structure de S engendrée par M.

EXEMPLES. Prenons pour Q le foncteur P_F d'oubli vers $\hat{M}$ de la catégorie $\hat{F}$ des foncteurs, pour H l'ensemble F des foncteurs associé à M_o, pour X l'ensemble des P_F-monomorphismes. La propriété (1) est vérifiée. Soient $C^\bullet \in \hat{F}_o$ et M une partie de C appartenant à $\tilde{M}_o$. La sous-catégorie $B^\bullet$ de $C^\bullet$ engendrée par M est une $(X.F_o,P_F)$-sous-structure de $C^\bullet$ engendrée par M. En effet, $B^\bullet$ est la P_F-sous-structure de $C^\bullet$ engendrée par M. D'après la proposition

2-3-II, $B^\bullet$ est un quotient de la catégorie libre $L[\hat{M}]$, où $[\hat{M}]$ est le sous-graphe de $[C^\bullet]$ engendré par M. Comme $\tilde{M}_o$ est un univers, on a

$$\hat{M} = M \cup \alpha(M) \cup \beta(M) \in \tilde{M}_o, \quad L[\hat{M}] \subset \bigcup_{n\in N} \hat{M}^n \in \tilde{M}_o ,$$

de sorte que le quotient B de $L[\hat{M}]$ appartient aussi à $\tilde{M}_o$. Donc P_F est $(M, X.F_o)$-engendrant. En utilisant ce résultat, on montre que les foncteurs d'oubli P_H relatifs aux "structures algébriques usuelles" (Chapitre II [4]) sont $(M, X.H_o)$-engendrants, si X est l'ensemble des P_H-monomorphismes.

3. *EXISTENCE DE LIMITES INDUCTIVES.*

Soient encore M et $\hat{M}$ les catégories d'applications associées aux univers M_o et $\hat{M}_o$ vérifiant

$$M_o \in \hat{M}_o \qquad \text{et} \qquad M_o \subset \hat{M}_o .$$

Supposons donné un foncteur Q de $\hat{H}^\bullet$ vers $\hat{M}$ et sa restriction q vers M d'une sous-catégorie pleine $H^\bullet$ de $\hat{H}^\bullet$ telle que H soit équipotent à un élément de $\hat{M}_o$.

PROPOSITION 4. *Supposons que Q soit à $\hat{M}_o$-produits et $(M, X.H^\bullet_o)$-engendrant, où X est un ensemble de Q-monomorphismes. Si tout couple de morphismes de $H^\bullet$ de même source et même but admet un noyau n dans q tel que $X.n \subset X$, alors $H^\bullet$ est une catégorie à $C^\bullet$-limites inductives, pour toute catégorie $C^\bullet$ vérifiant $C \in \hat{M}_o$.*

Δ. Soit F un foncteur de $C^\bullet$ vers $H^\bullet$. Considérons l'ensemble I de toutes les transformations naturelles $i = (\hat{s}_i, t_i, F)$ de F vers un foncteur $\hat{s}_i$ constant sur $s_i \in H^\bullet_o$. La surjection $i \to (t_i(c))_{c\in C_o}$ définit une bijec-

-tion de I sur une partie de $H^{C_o^\bullet}$. Comme H et $C_o^\bullet$ sont équipotents à des éléments de $\mathcal{M}_o$, ce produit l'est aussi, et il existe un produit naturalisé $((p_i)_{i\in I},S)$ de $(s_i)_{i\in I}$ dans Q. Si Φ est un foncteur de $C^\bullet$ vers $H^\bullet$ (resp. vers $\mathcal{M}$), notons $\bar{\Phi}$ le foncteur de $C^\bullet$ vers $\hat{H}^\bullet$ (resp. vers $\hat{\mathcal{M}}$) défini par la surjection $\underline{\Phi}$; nous utilisons une notation analogue pour les transformations naturelles. - Le foncteur q.F admet une limite inductive naturalisée canonique σ , car $\mathcal{M}$ est une catégorie à $C^\bullet$-limites inductives; désignons par σ_c l'injection canonique de q.F(c) dans $M = \varinjlim$ q.F. Par construction de σ, la transformation naturelle $\bar{\sigma}$ est une limite inductive naturalisée du foncteur $\overline{q.F}$ de $C^\bullet$ vers $\hat{\mathcal{M}}$.

- Pour tout $c \in C_o^\bullet$, il existe un crochet $t(c) = [t_i(c)]_{i\in I}$ de source F(c), de but S . Si $m \in c'.C.c$, on trouve

$t(c').F(m)=[t_i(c')]_{i\in I}.F(m)=[t_i(c').F(m)]_{i\in I}=[t_i(c)]_{i\in I}=t(c)$,

de sorte que $T = (\hat{S},t,\bar{F})$ est une transformation naturelle et $p_i T = \bar{I}$ pour tout $i \in I$. Soit k l'application $\varinjlim^{\bar{\sigma}} QT$ de M dans Q(S) telle que $k\bar{\sigma} = QT$. Le foncteur Q étant $(\mathcal{M},X.H_o^\bullet)$-engendrant et M appartenant à $\mathcal{M}_o$, il existe un $(X.H_o^\bullet,Q)$-sous-morphisme j de S engendré par k. Montrons que $s=\alpha(j)$ est une limite inductive de F. Or il existe k' tel que $Q(j).k' = k$; il s'ensuit

$$Q(t(c)) = k.\sigma_c = Q(j).k'.\sigma_c$$

et, j étant un Q-monomorphisme, il existe un unique t'(c) vérifiant

$$Q(t'(c)) = k'.\sigma_c \quad \text{et} \quad t(c) = j.t'(c),$$

ce pour tout $c \in C_o^\bullet$. Si $m \in c'.C.c$, on obtient

$$j.t'(c').F(m) = t(c').F(m) = t(c) = j.t'(c),$$

d'où $t'(c').F(m) = t'(c)$, car j est un monomorphisme. Ceci montre que $T' = (\hat{s},t',F)$ est une transformation natu-

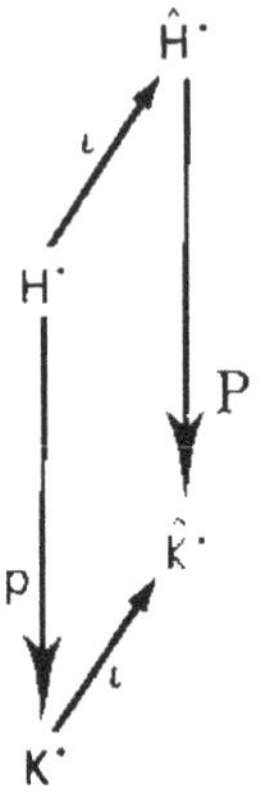

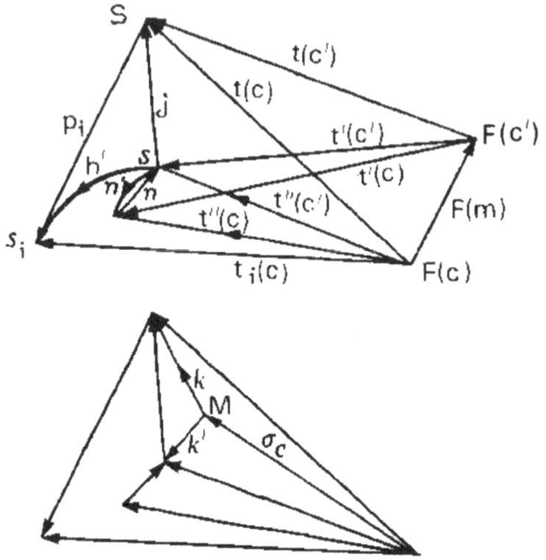

relle, qui satisfait les égalités

$$Q\bar{T}' = k'\bar{\sigma} \quad \text{et} \quad j\bar{T}' = T.$$

Montrons que T' est une limite inductive naturalisée. Pour cela, soit $i = (\hat{s}_i, t_i, F) \in I$. En posant $h = p_i.j$, on a

$$h\bar{T}' = p_i(j\bar{T}') = p_i T = \bar{I}\ , \text{ d'où } hT' = i.$$

Supposons aussi $h'T' = i$. Par hypothèse, il existe un noyau n de (h',h) dans q avec $X.n \subset X$. Soit $c \in C_o^{\bullet}$. Comme

$$h.t'(c) = t_i(c) = h'.t'(c),$$

il existe un unique $t''(c)$ vérifiant $n.t''(c) = t'(c)$. Si $m \in c'.C.c$, on obtient

$$n.t''(c').F(m) = t'(c').F(m) = t'(c) = n.t''(c).$$

Il s'ensuit, n étant un monomorphisme, $t''(c').F(m) = t''(c)$. Par suite $T'' = (\widehat{\alpha(n)}, t'', F)$ est une transformation naturelle, et $nT'' = T'$. Si $k'' = \lim^{\bar{\sigma}} Q\bar{T}''$, on trouve

$$(Q(j.n).k'')\bar{\sigma} = Q(j.n)Q\bar{T}'' = Q(j(n\bar{T}'')) = Q(j\bar{T}') = QT = k\bar{\sigma}$$

et, $\bar{\sigma}$ étant une limite inductive naturalisée, $Q(j.n).k'' = k$. Par définition de j, il existe n' tel que $(j.n).n' = j$. Cette relation a pour conséquence $n' = n^{-1}$, car j et n

sont des monomorphismes. Par suite $h = h'$, et T' est une limite inductive naturalisée de F. ∇

REMARQUES. 1° On pourrait déduire la proposition 4 de la proposition 1, en prenant pour p le foncteur canonique de $H^{\cdot}$ vers la catégorie $\mathbf{n}(H^{\cdot},C^{\cdot})^{\square}$ de transformations naturelles; dans ce cas, il faut montrer que les conditions de la proposition 1 sont satisfaites, ce qui est aussi difficile que de prouver directement la proposition 4.

2° Avec les hypothèses de la proposition 4, et en supposant que $C^{\cdot}$ est une catégorie discrète, on voit que $H^{\cdot}$ est une catégorie à M_o-sommes. La proposition 2 affirmant que, sous ces hypothèses, q est un foncteur à structures quasi-quotients, on déduit de cette seule assertion que $H^{\cdot}$ est à $C^{\cdot}$-limites inductives pour toute catégorie $C^{\cdot}$ vérifiant $C \in M_o$, à l'aide de la proposition 6-5-I [4]. Mais la démonstration de la proposition 4 ne se simplifie guère en y supposant $C^{\cdot}$ discrète.

4. COMPLETION D'UNE CATEGORIE.

Soient M et $\bar{M}$ les catégories d'applications considérées dans les deux § précédents. Nous supposons donnés deux ensembles I et J de catégories $I^{\cdot}$ telles que $I \in M_o$.

Désignons par S_o^{IJ} l'ensemble des triplets $u = (C^{\cdot},\mu,\nu)$, où $C^{\cdot}$ est un graphe multiplicatif, où $C \in M_o$ et où μ (resp. où ν) est une surjection associant à certains néofoncteurs F de $I^{\cdot} \in I$ (resp. $I^{\cdot} \in J$) vers $C^{\cdot}$ une transformation naturelle de la forme $\mu(F) = (F,t,\hat{e})$ (resp. $\nu(F) = (\hat{e},t,F)$), en notant $\hat{e}$ le néofoncteur de $I^{\cdot}$ vers $C^{\cdot}$ constant sur $e \in C^\bullet_o$. Soit S^{IJ} l'ensemble des triplets (u',Φ,u) où $u = (C^\bullet,\mu,\nu) \in S^{IJ}_o$, $u' = (C'^\bullet,\mu',\nu') \in S^{IJ}_o$ et où Φ est un néofoncteur de $C^\bullet$ vers $C'^\bullet$ vérifiant la condition: Si $\mu(F)$ est défini, $\mu'(\Phi.F)$ est défini et égal à $\Phi\mu(F)$; si $\nu(F')$ est défini, $\nu'(\Phi.F')$ est défini et égal à $\Phi\nu(F')$.

S^{IJ} devient une catégorie pour la loi de composition:

$$(u'',\Phi',u'_1).(u',\Phi, u) = (u'',\Phi'.\Phi,u)$$

si, et seulement si, $u'_1 = u'$. Nous identifions S^{IJ}_o à la classe de ses unités. La surjection

$$(u',\Phi,u) \to p_{N'}(\Phi) = (C',\underline{\Phi},C)$$

définit un foncteur p''^{IJ} de S^{IJ} vers M. Soit F'^{IJ} la sous-catégorie pleine de S^{IJ} ayant pour unités les triplets $u = (C^\bullet,\mu,\nu)$ tels que $C^\bullet$ soit une catégorie et que μ soit une application I-limite projective naturalisée partielle, ν une application J-limite inductive naturalisée partielle, sur $C^\bullet$. Soit F^{IJ} la sous-catégorie pleine de F'^{IJ} ayant pour unités les triplets $u = (C^\bullet,\mu,\nu) \in F'^{IJ}_o$ tels que μ et ν soient des applications I-limite projective naturalisée et J-limite inductive naturalisée respectivement, sur $C^\bullet$. Nous notons p'^{IJ} et p^{IJ} les foncteurs de F'^{IJ} et de F^{IJ} respectivement vers M restrictions de p''^{IJ}.

De même nous définissons à partir de l'univers $\hat{M}_o$ les catégories $\hat{S}^{IJ}$, $\hat{F}'^{IJ}$ et $\hat{F}^{IJ}$, les foncteurs P''^{IJ}, P'^{IJ} et P^{IJ} (toujours relatives aux mêmes ensembles I et J).

PROPOSITION 5. Les foncteurs p^{IJ}, p'^{IJ} *et* p''^{IJ} *sont à* M_o*-produits, ainsi que les foncteurs injections canoniques de* F^{IJ} *vers* F'^{IJ} *et de* F'^{IJ} *vers* S^{IJ}.

Δ. Supposons $u_d = (C^\bullet_d,\mu_d,\nu_d) \in S^{IJ}_o$ pour tout $d \in D$, où $D \in M_o$. Soit $C^\bullet$ le graphe multiplicatif produit $\prod_{d\in D} C^\bullet_d$

et p_d le néofoncteur projection canonique de $C^\cdot$ sur $C^\cdot_d$. Soit $\alpha(\mu)$ l'ensemble des néofoncteurs F de $I^\cdot \in \mathcal{I}$ vers $C^\cdot$ tels que $\mu_d(p_d.F) = (F_d,t_d,\hat{e}_d)$ soit défini pour tout $d \in D$. Si $i \in I^\cdot_o$, notons $t(i)$ la famille $(t_d(i))_{d\in D}$; le triplet $(F,t,\hat{e})$, où $e = (e_d)_{d\in D}$, est une transformation naturelle $\mu(F)$. De même, si F' est un foncteur de $J^\cdot \in \mathcal{J}$ vers $C^\cdot$ et si $\nu_d(p_d.F') = (\hat{e}'_d,t'_d,F'_d)$ est défini pour tout $d \in D$, nous désignons par $\nu(F')$ la transformation naturelle $(\hat{e}',t',F')$ définie par

$$e' = (e'_d)_{d\in D} \quad \text{et} \quad t'(i) = (t'_d(i))_{d\in D} \text{ pour } i \in I^\cdot_o.$$

Nous obtenons ainsi deux surjections μ et ν et $u = (C^\cdot,\mu,\nu)$ est un élément de $\mathcal{S}^{\mathcal{I}\mathcal{J}}_o$. Montrons que u est un produit de $(u_d)_{d\in D}$ dans $\mathcal{S}^{\mathcal{I}\mathcal{J}}$. En effet, par construction, on a

$$\Pi_d = (u_d,p_d,u) \in \mathcal{S}^{\mathcal{I}\mathcal{J}} \quad \text{pour tout } d \in D.$$

Supposons $\Phi_d = (u_d,\phi_d,u') \in \mathcal{S}^{\mathcal{I}\mathcal{J}}$ pour tout $d \in D$, où $u' = (C'^\cdot,\mu',\nu')$; soit $\phi = [\phi_d]_{d\in D}$ le néofoncteur canonique de $C'^\cdot$ vers $C^\cdot$. Si $\mu'(F') = (F',t',\hat{e}')$ est défini, $\mu_d(\phi_d.F')$ est défini pour tout $d \in D$. Comme $\phi_d.F' = p_d.\phi.F'$, la transformation naturelle $\mu(\phi.F')$ est définie et égale à

$$(\phi.F',t,\hat{e}), \quad \text{si} \quad t(i) = (\phi_d(t'(i)))_{d\in D} = \phi(t'(i))$$

pour tout $i \in I^\cdot_o$, c'est-à-dire à $\phi\mu'(F')$. On voit de même que, si $\nu'(F'')$ est défini, $\nu(\phi.F'')$ est défini et égal à $\phi\nu'(F'')$. Par suite $\Phi = (u,\phi,u')$ appartient à $\mathcal{S}^{\mathcal{I}\mathcal{J}}$ et l'on a $\Phi_d = p_d.\Phi$ pour tout $d \in D$. Ceci prouve que le foncteur $p''^{\mathcal{I}\mathcal{J}}$ est à $\mathcal{M}_o$-produits.

- Supposons de plus $u_d \in \mathcal{F}'^{\mathcal{I}\mathcal{J}}$ pour tout $d \in D$. Avec les notations précédentes, $C^\cdot$ est une catégorie. Montrons que, si $\mu(F) = (F,t,\hat{e})$ est défini, c'est une limite projective naturalisée. En effet, soit $\psi = (F,t',\hat{e}')$ une transformation naturelle. Etant donné que $\mu_d(p_d.F)$ est une limite projective naturalisée et que $p_d\psi$ est une transformation naturelle vers $p_d.F$, il existe $k_d = \varprojlim p_d\psi$ tel que

$$\mu_d(p_d.F)\ k_d = p_d\psi\ ;$$

en posant $k = [k_d]_{d\varepsilon D}$, on voit que k est l'unique élément de C vérifiant $\mu(F)\ k = \psi$. Donc $\mu(F)$ est une limite projective naturalisée. On montre d'une manière analogue que ν est une application $\mathcal{J}$-limite inductive naturalisée partiel-le sur $C^\bullet$. Il en résulte $u \in \mathcal{F}'^{IJ}$, et u est aussi le produit de $(u_d)_{d\varepsilon D}$ dans $\mathcal{F}'^{IJ}$. Ainsi p'^{IJ} et $(\mathcal{S}^{IJ}, \iota, \mathcal{F}'^{IJ})$ sont des foncteurs à $\mathcal{M}_o$-produits.

- Enfin si $u_d \in \mathcal{F}_o^{IJ}$ pour tout $d \in D$, la limite $\mu(F)$ est définie pour tout foncteur F de $I^\bullet \in \mathcal{I}$ vers $C^\bullet$ et $\nu(F')$ est définie pour tout foncteur F' de $J^\bullet \in \mathcal{J}$ vers $C^\bullet$. Par suite u appartient à $\mathcal{F}^{IJ}$ et c'est le produit de $(u_d)_{d\varepsilon D}$ dans $\mathcal{F}^{IJ}$. ∇

PROPOSITION 6. Les foncteurs p^{IJ}, p'^{IJ}, p''^{IJ} sont à noyaux, de même que les foncteurs injections canoniques de $\mathcal{F}^{IJ}$ vers $\mathcal{F}'^{IJ}$ et de $\mathcal{F}'^{IJ}$ vers $\mathcal{S}^{IJ}$.

Δ. Supposons que $\Phi = (\bar{u},\phi,u)$ et $\Phi' = (\bar{u},\phi',u)$ soient deux éléments de $\mathcal{S}'^{IJ}$, où $u = (C^\bullet,\mu,\nu)$ et $\bar{u} = (\bar{C}^\bullet,\bar{\mu},\bar{\nu})$. Soit $G^\bullet$ le graphe multiplicatif noyau de (ϕ,ϕ'), qui est formé des $x \in C$ tels que $\phi(x) = \phi'(x)$; soit η le néofoncteur injection canonique de $G^\bullet$ vers $C^\bullet$. Soit F' un néofoncteur de $I^\bullet \in \mathcal{I}$ vers $G^\bullet$, et F le néofoncteur $\eta.F'$. Supposons $\mu(F) = (F,t,\hat{e})$ défini; alors

$$\bar{\mu}(\phi.F) = \phi\mu(F) \quad \text{et} \quad \bar{\mu}(\phi'.F) = \phi'\mu(F).$$

Les égalités

$$\phi.F = \phi.\eta.F' = \phi'.\eta.F' = \phi'.F$$

entraînent $\phi\mu(F) = \phi'\mu(F)$, d'où $t(I_o^\bullet) \subset G$. Il s'ensuit que

$(F', t, \hat{e})$ est aussi une transformation naturelle, que nous notons $\mu'(F')$; on a $\eta\mu'(F') = \mu(F)$. De même si $\nu(F)$ est défini et égal à $(\hat{e}', t', F)$, on voit que $(\hat{e}', t', F')$ est une transformation naturelle, désignée par $\nu'(F')$. De cette façon, on construit deux surjections μ' et ν', et

$$u' = (G^{\bullet}, \mu', \nu') \in \mathcal{S}_o^{IJ}, \quad \bar{\eta} = (u, \eta, u') \in \mathcal{S}^{IJ}.$$

Supposons $\Phi'' = (u, \phi'', u'') \in \mathcal{S}^{IJ}$, où

$$u'' = (C''^{\bullet}, \mu'', \nu'') \quad \text{et} \quad \Phi.\Phi'' = \Phi'.\Phi''.$$

La surjection $\underline{\phi}''$ définit un néofoncteur ϕ''_1 de $C''^{\bullet}$ vers $G^{\bullet}$ vérifiant $\eta.\phi''_1 = \phi''$. Soit F'' un foncteur de $I^{\bullet} \in I$ vers $C''^{\bullet}$ tel que $\mu''(F'')$ soit défini, et posons

$$F' = \phi''_1.F'' \quad \text{et} \quad F = \phi''.F'' ;$$

alors $F = \eta.F'$. Puisque Φ'' appartient à $\mathcal{S}^{IJ}$, la transformation naturelle $\mu(F)$ est définie et égale à $\phi''\mu''(F'')$. On en déduit que $\mu'(F')$ est défini et est égal à $\phi''_1\mu''(F'')$. D'une manière analogue, si $\nu''(F''_1)$ est défini, alors $\nu'(\phi''_1.F''_1)$ est défini et vaut $\phi''_1\nu''(F''_1)$. Par conséquent (u', ϕ''_1, u'') appartient à $\mathcal{S}^{IJ}$, et c'est l'unique élément Φ''_1 de $\mathcal{S}^{IJ}$ pour lequel $\bar{\eta}.\Phi''_1 = \Phi''$. Ceci signifie que $\bar{\eta}$ est un noyau de (Φ, Φ') dans p''^{IJ}.

- Supposons de plus que Φ et Φ' appartiennent à $\mathcal{F}'^{IJ}$, et montrons que $u' \in \mathcal{F}'^{IJ}$. En effet, il suffit de montrer que si $\mu'(F')$ est défini c'est une limite projective naturalisée. Or soit ψ une transformation naturelle $(F', \tau, \hat{e}_1)$, et $F = \eta.F'$. Etant donné que ψ est une transformation naturelle et que $\mu(F) = \eta\mu'(F')$ est une limite projective naturalisée, il existe k tel que $\mu(F)k = \eta\psi$. Les égalités

$$\bar{\mu}(\phi.F)\phi(k) = \phi(\mu(F)k) = (\phi.\eta)\psi = (\phi'.\eta)\psi =$$
$$= \phi'(\mu(F)k) = \bar{\mu}(\phi'.F)\phi'(k)$$

ont pour conséquence $\phi(k) = \varprojlim (\phi.\eta)\psi = \phi'(k)$, d'où $k \in G$

et $\mu'(F')k = \psi$. Ainsi $\mu'(F')$ est une limite projective naturalisée. On prouve de même que $\nu'(F'')$ est une limite inductive naturalisée lorsque cette transformation naturelle est définie. Donc $u' \in F'^{IJ}$ et $\bar{\eta}$ est un noyau de (Φ,Φ') dans F'^{IJ}. - Enfin lorsque Φ et Φ' appartiennent à F^{IJ}, la construction de u' entraîne que c'est un élément de F^{IJ}, de sorte que $\bar{\eta}$ est un noyau de (Φ,Φ') dans p^{IJ}. ∇

Définition. Soient $C^\cdot$ une catégorie et μ une application I-limite projective naturalisée partielle sur $C^\cdot$. On dit qu'une partie M de C est *saturée pour* μ si elle vérifie la condition: Lorsque $\mu(F) = (F,t,\hat{e})$ est défini pour un foncteur F de $I^\cdot$ vers $C^\cdot$ et que $F(I) \subset M$, on a

$$t(I_o^\cdot) \subset M \quad \text{et} \quad \lim^{\mu(F)}\psi \in M$$

pour toute transformation naturelle $\psi = (F,t',\hat{e}')$ vérifiant $t'(I_o^\cdot) \subset M$. On définit d'une façon analogue la notion d'une partie de C *saturée pour* ν, si ν est une application J-limite inductive naturalisée sur $C^\cdot$ (partielle).

Soit X' l'ensemble des

$$\Phi = ((C^\cdot,\mu,\nu),\phi,(G^\cdot,\mu',\nu')) \in \hat{F}'^{IJ}$$

tels que ϕ soit un foncteur injectif, que $\mu'(F')$ soit défini si, et seulement si, $\mu(\phi.F')$ est défini et que $\nu'(F')$ soit défini dans le seul cas où $\nu(\phi.F')$ est défini.

PROPOSITION 7. Soit $u = (C^\cdot,\mu,\nu) \in F'^{IJ}$ *et* ϕ *un foncteur de* $G^\cdot$ *vers* $C^\cdot$. *On a* $P_F(\phi) \in P'^{IJ^o}(u.X')$ *si, et seulement si,* ϕ *est un foncteur injectif et si* $\phi(G)$ *est saturé pour* μ *et pour* ν. *Tout élément de* X' *est un* P''^{IJ}*-monomorphisme.*

Δ. Supposons

$$\Phi = ((C^{\bullet},\mu,\nu),\phi,(G^{\bullet},\mu',\nu')) \in X';$$

alors ϕ est injectif et $M = \phi(G)$ définit une sous-catégorie $M^{\bullet}$ de $C^{\bullet}$. Soit ξ l'isomorphisme de $M^{\bullet}$ sur $G^{\bullet}$ défini par la bijection $\underline{\phi}^{-1}$. Supposons $\mu(F) = (F,t,\hat{e})$ défini pour un foncteur F de $I^{\bullet} \in \mathcal{I}$ vers $C^{\bullet}$ tel que $F(I) \subset M$. Comme $F = \phi.F'$, où $F' = \xi.(M^{\bullet},\underline{F},I^{\bullet})$, par définition de X' la limite projective naturalisée $\mu'(F')$ est définie. L'égalité $\mu(F) = \phi\mu'(F')$ prouve que

$$\mu'(F') = (F',\underline{\xi}t,\hat{s}) \,, \quad \text{où} \quad e = \phi(s);$$

il s'ensuit

$$t(I_o^{\bullet}) = \phi(\underline{\xi}t(I_o^{\bullet})) \subset \phi(G) = M \,.$$

Soit ψ une transformation naturelle $(F,t',\hat{e}')$ vérifiant $t'(I_o^{\bullet}) \subset M$; si ψ' est la transformation naturelle

$$(F',\underline{\xi}t',\hat{s}'), \quad \text{où} \quad \phi(s') = e',$$

il existe $k' = \underleftarrow{\lim}^{\mu'(F')}\psi' \in G$. Les relations

$$\phi\psi' = \psi \quad \text{et} \quad \phi\mu'(F') = \mu(F)$$

entraînent

$$\underleftarrow{\lim}^{\mu(F)}\psi = \underleftarrow{\lim}^{\mu(F)}\phi\psi' = \phi(\underleftarrow{\lim}^{\mu'(F')}\psi') = \phi(k') \in M.$$

Donc M est saturé pour μ. On montre de même que M est saturé pour ν.

- Avec les notations précédentes, prouvons que Φ est un $P''^{\mathcal{I}\mathcal{J}}$ - monomorphisme. Pour cela, soit

$$\Phi' = ((C^{\bullet},\mu,\nu),\phi',(K^{\bullet},\hat{\mu},\hat{\nu})) \in \hat{\mathcal{S}}^{\mathcal{I}\mathcal{J}} \quad \text{et} \quad P_{\mathcal{F}}(\phi') = P_{\mathcal{F}}(\phi).f''.$$

D'après la proposition 4-2-II [4], f'' définit un néofoncteur ϕ'' de $K^{\bullet}$ vers $G^{\bullet}$. Si $\hat{\mu}(F_1) = (F_1,t_1,\hat{e}_1)$ est défini, $\mu(\phi'.F_1)$ est définie et l'égalité $\phi'.F_1 = \phi.(\phi''.F_1)$ assure que $\mu'(\phi''.F_1)$ est aussi défini. Puisque

$$\phi\mu'(\phi''.F_1) = \mu(\phi'.F_1) = \phi'\hat{\mu}(F_1) = \phi(\phi''\hat{\mu}(F_1))$$

et que ϕ est injectif, on trouve $\mu'(\phi''.F_1) = \phi''\hat{\mu}(F_1)$. D'une manière analogue, si $\hat{\nu}(F'_1)$ est défini, $\nu'(\phi''.F'_1)$ est défini et égal à $\phi''\ \hat{\nu}(F'_1)$. Par suite

$$((G^\bullet,\mu',\nu'),\phi'',(K^\bullet,\hat{\mu},\hat{\nu}))$$

est l'unique élément Φ'' de $\hat{S}^{IJ}$ tel que

$$\Phi.\Phi'' = \Phi' \text{ et } \quad P_F(\phi'') = f''.$$

Autrement dit, Φ est un P''^{IJ}-monomorphisme.

- Inversement, supposons que ϕ soit un foncteur injectif de $G^\bullet$ vers $C^\bullet$ et que $\phi(G) = M$ soit saturé pour μ et pour ν. L'ensemble M définit une sous-catégorie de $C^\bullet$. Si F est un foncteur de $I^\bullet \varepsilon\ I$ vers $C^\bullet$ tel que $F(I) \subset M$ et que $\mu(F) = (F,t,\hat{e})$ soit défini, et si F_1 est le foncteur de $I^\bullet$ vers $M^\bullet$ défini par $\underline{F}$, le triplet $(F_1,t,\hat{e})$ est une limite projective naturalisée ψ, car M est saturé pour μ. En notant ξ l'isomorphisme de $M^\bullet$ sur $G^\bullet$ défini par $\underline{\phi}^{-1}$, la transformation naturelle $\xi\psi$ est aussi une limite projective naturalisée, que nous noterons $\mu'(\phi.F_1)$. Soit μ' la surjection associant, à tout foncteur F' de $I^\bullet\ \varepsilon\ I$ vers $G^\bullet$ tel que $\mu(\phi.F')$ soit défini, la limite projective naturalisée $\mu'(\xi.(\phi.F')_1) = \mu'(F')$. De même, si F'' est un foncteur de $I^\bullet\ \varepsilon\ J$ vers $G^\bullet$ et si $\nu(\phi.F'')$ est défini, notons $\nu'(F'')$ l'unique limite inductive naturalisée vérifiant l'égalité $\phi\nu(F'') = \nu(\phi.F'')$. Il est évident que

$$((C^\bullet,\mu,\nu),\phi,(G^\bullet,\mu',\nu')) \ \varepsilon \ X',$$

et c'est l'unique élément de $u.X'$ appliqué par P'^{IJ} sur $P_F(\phi)$, ce qui achève la démonstration. ∇

COROLLAIRE. $X = X' \cap \hat{F}^{IJ}$ est l'ensemble des P^{IJ}-monomorphismes.

Δ. Soit Φ un P^{IJ}-monomorphisme $((C^\bullet,\mu,\nu),\phi,(G^\bullet,\mu',\nu'))$; ϕ est un foncteur injectif. Soit F' un foncteur de $I^\bullet\ \varepsilon\ I$

vers $G^\bullet$; comme μ et μ' sont des applications I-limite projective naturalisées, $\mu(\phi.F')$ et $\mu'(F')$ sont définis et, Φ appartenant à $\hat{F}^{IJ}$, on a $\mu(\phi.F') = \phi\mu'(F')$. De même $\nu'(F'')$ est l'unique transformation naturelle vérifiant $\nu(\phi.F'') = \phi\nu'(F'')$ pour tout foncteur F'' de $I^\bullet \varepsilon J$ vers $G^\bullet$. Donc $\Phi \varepsilon X$. - Inversement, tout élément de X est un P^{IJ}-monomorphisme, car c'est un P'^{IJ}-monomorphisme (proposition 7) et $\hat{F}^{IJ}$ est une sous-catégorie pleine de $\hat{F}'^{IJ}$. ∇

PROPOSITION 8. Si I et J sont équipotents à des éléments de M_o, le foncteur P'^{IJ} est $(M, X'.F'^{IJ}_o)$-engendrant et P^{IJ} est $(M, X.F^{IJ}_o)$-engendrant.

Δ. Soit $u = (C^\bullet, \mu, \nu) \varepsilon \hat{F}'^{IJ}_o$ et soit M une partie de C appartenant à la saturante $\bar{M}_o$ de M_o dans $\hat{M}_o$. Soit B l'ensemble des parties B de C définissant une sous-catégorie de $C^\bullet$ et saturées pour μ et ν. Il est clair que l'intersection G de B appartient à B. Le foncteur P'^{IJ} vérifiant la propriété (1) du § 3, pour prouver que P'^{IJ} est $(M, X'.F'^{IJ}_o)$-engendrant, il suffit de montrer qu'il existe une $(X'.F'^{IJ}_o, P'^{IJ})$-sous-structure de u engendrée par M. Or, d'après la proposition 7, si une telle sous-structure existe, elle est de la forme $(G^\bullet, \mu', \nu')$, et G appartient à $\bar{M}_o$. Nous sommes donc ramenés à montrer que G est équipotent à un élément de M_o. Pour cela, nous allons construire G par récurrence transfinie.

- Nous aurons à utiliser quelques notions sur les (nombres) ordinaux que nous allons rappeler. Soit λ un ordinal et O_λ l'ensemble bien ordonné des ordinaux inférieurs à λ. Si λ n'a pas de prédécesseur dans O_λ, on l'appelle un ordinal limite. On dit que λ est un ordinal régulier si, dans O_λ, on a $\sup_{\xi<\lambda'} \lambda_\xi < \lambda$ pour toute suite transfinie $(\lambda_\xi)_{\xi<\lambda'}$ telle que

$$\lambda_\xi < \lambda \quad \text{et} \quad \lambda' < \lambda \ .$$

A tout ensemble A, nous associons l'ordinal $\bar{A}$, qui est le plus petit ordinal équipotent à A; un tel ordinal est appelé ordinal initial. Il existe une bijection γ du sous-ensemble bien ordonné de $O_{\bar{A}}$ formé des ordinaux initiaux inférieurs à $\bar{A}$ sur une section commençante de $O_{\bar{A}}$; l'ordinal $\gamma^{-1}(\lambda)$ est dit ordinal initial d'indice λ, et noté ω_λ. Si λ admet un prédécesseur, ω_λ est régulier. Si ω_λ est régulier tandis que λ est un ordinal limite, on dit que ω_λ est inaccessible; dans ce cas, on montre que $\lambda = \omega_\lambda$. Soit Λ l'ordinal borne supérieure des ordinaux $\bar{A}$, où $A \in M_o$; on a aussi $\Lambda = \sup_{A\in M_o} \bar{A}$. Comme M_o est un univers, on prouve que Λ est un ordinal inaccessible. Si $I^\bullet \in I \cup J$, on a $\bar{I} < \Lambda$; soit Λ' l'ordinal borne supérieure des ordinaux $\bar{I}$, où $I^\bullet$ décrit $I \cup J$. L'hypothèse $I \in \tilde{M}_o$ et $J \in \tilde{M}_o$ entraînant $\overline{I \cup J} < \Lambda$ et Λ étant régulier, Λ' est strictement inférieur à Λ; a fortiori $\Lambda'+1 < \Lambda$, d'où $\omega_{\Lambda'+1} < \omega_\Lambda = \Lambda$. Notons $\hat{\lambda}$ l'ordinal initial régulier $\omega_{\Lambda'+1}$.

- Soit $M_1^\bullet$ la sous-catégorie de $C^\bullet$ engendrée par M. Nous avons vu (fin § 3) que M_1 appartient à $\tilde{M}_o$. Soit λ un ordinal inférieur ou égal à $\hat{\lambda}$ et supposons définie une suite transfinie $(M_\xi)_{\xi<\lambda}$ vérifiant la condition:

(σ) M_ξ appartient à $\tilde{M}_o$ et définit une sous-catégorie de $C^\bullet$; on a $M_{\xi'} \subset M_\xi$ lorsque $\xi' < \xi < \lambda$.

Définissons M_λ comme suit:

1er cas: Si λ est un ordinal limite, on pose $M_\lambda = \bigcup_{\xi<\lambda} M_\xi$. Evidemment M_λ définit une sous-catégorie de $C^\bullet$ contenant

M_ξ et, λ étant l'ordinal associé à un élément de $\mathcal{M}_o$, on a $M_\lambda \in \tilde{\mathcal{M}}_o$.

$2^{\text{ème}}$ cas: $\lambda = \xi + 1$. Si F est un foncteur de $I^\bullet$ vers $C^\bullet$ tel que $F(I) \subset M_\xi$ et que $\mu(F) = (F,t,\hat{e})$ soit défini, notons A_F l'ensemble formé des éléments $t(i)$, où $i \in I^\bullet_o$, et des éléments $h = \underleftarrow{\lim}^{\mu(F)}\psi$, pour toute transformation naturelle $\psi = (F,t_1,\hat{e}_1)$ telle que $t_1(I^\bullet_o) \subset M_\xi$. La surjection telle que

$$i \to t(i), \quad (t_1(i))_{i\in I^\bullet_o} \to h$$

applique une partie U_F de $I \cup (M_\xi)^{I^\bullet_o}$ sur A_F. Puisque I et M_ξ appartiennent à $\tilde{\mathcal{M}}_o$, on a $U_F \in \tilde{\mathcal{M}}_o$, d'où $A_F \in \tilde{\mathcal{M}}_o$. Si F' est un foncteur de $J^\bullet \in \mathcal{J}$ vers $C^\bullet$ tel que l'on ait $F'(J) \subset M_\xi$ et que $\nu(F') = (\hat{e}',t',F')$ soit défini, notons $A'_{F'}$ l'ensemble formé des éléments $t'(i)$, où $i \in J^\bullet_o$, et des éléments $\underrightarrow{\lim}^{\nu(F')}\psi'$, pour toute transformation naturelle $\psi' = (\hat{e}'',t'_1,F')$ vérifiant $t'_1(J^\bullet_o) \subset M_\xi$; on trouve encore $A'_{F'} \in \tilde{\mathcal{M}}_o$. Soit M'_ξ l'ensemble réunion des ensembles A_F, où $F \in \alpha(\mu)$, et des ensembles $A'_{F'}$, où $F' \in \alpha(\nu)$. Nous noterons $M^\bullet_\lambda$ la sous-catégorie de $C^\bullet$ engendrée par M'_ξ. La bijection $F \to (F(i))_{i\in I}$ appliquant $\alpha(\mu)$ sur une partie de $\bigcup_{I'\in\mathcal{I}}(M_\xi)^I \in \tilde{\mathcal{M}}_o$, on a $\overline{\overline{\alpha(\mu)}} < \Lambda$. Il en résulte que l'ensemble réunion M'_ξ appartient à $\tilde{\mathcal{M}}_o$, car, de même, $\overline{\overline{\alpha(\nu)}} < \Lambda$. A fortiori, $M_\lambda \in \tilde{\mathcal{M}}_o$.

- Dans les deux cas, nous avons ainsi construit une suite transfinie $(M_\zeta)_{\zeta\leq\lambda}$ vérifiant (σ). Par récurrence transfinie, on obtient de cette façon une suite transfinie croissante $(M^\bullet_\xi)_{\xi\leq\hat{\lambda}}$ de sous-catégories de $C^\bullet$ telle que $M_{\hat{\lambda}} \in \tilde{\mathcal{M}}_o$. Posons $A = M_{\hat{\lambda}}$; par définition de A, on voit que A est contenu dans tout élément de $\mathcal{B}$. Montrons que A est saturé pour μ et pour ν, d'où il résultera que A est l'intersection G de $\mathcal{B}$. En effet, supposons $\mu(F) = (F,t,\hat{e})$ défini pour un foncteur F de $I^{\bullet}$ vers $C^{\bullet}$ tel que $F(I) \subset A$. Pour tout $i \in I$, il existe un $\xi_i < \hat{\lambda}$ tel que $F(i) \in M_{\xi_i}$; étant donné que $I \in \mathcal{M}_o$ entraîne $\bar{I} < \hat{\lambda}$ et que $\hat{\lambda}$ est un ordinal régulier, on a $\xi = \sup_{i \in I} \xi_i < \hat{\lambda}$. Par suite $F(I) \subset M_\xi$ et, par construction de $M_{\xi+1}$, on obtient

$$t(I_o^{\bullet}) \subset M'_\xi \subset M_{\xi+1} \subset A .$$

De plus, si ψ est une transformation naturelle $(F,t',\hat{e}')$ telle que $t'(I_o^{\bullet}) \subset A$, il existe pour la même raison un ordinal ξ' vérifiant $\xi < \xi' < \hat{\lambda}$ tel que

$$t'(I_o^{\bullet}) \subset M_{\xi'} \text{ , d'où } \underleftarrow{\lim}^{\mu(F)} \psi \in M'_{\xi'} \subset A.$$

Ceci prouve que A est saturé pour μ. Un raisonnement analogue montre que A est saturé pour ν. Donc $A \in \mathcal{B}$, de sorte que $G = A$, c'est-à-dire que G appartient à $\tilde{\mathcal{M}}_o$. Ainsi P'^{IJ} est $(\mathcal{M},X'.F_o'^{IJ})$-engendrant.

- Si l'on suppose dans ce qui précède que u appartient à $\bar{F}_o^{IJ}$, le corollaire de la proposition 7 entraîne que G définit également une $(X.F_o^{IJ},P^{IJ})$-sous-structure $(G^{\bullet},\mu',\nu')$ de u engendrée par M. Par suite P^{IJ} est $(\mathcal{M},X.F_o^{IJ})$-engendrant, ce qui achève la démonstration. ∇

PROPOSITION 9. Si I *et* J *appartiennent à* $\tilde{\mathcal{M}}_o$ *, la catégorie* S^{IJ} *est à* F'^{IJ}*-projections et à* F^{IJ}*-projections; la catégorie* F'^{IJ} *est à* F^{IJ}*-projections.*

Δ. Nous nous ramènerons au théorème général d'existence de structures libres. Pour cela, montrons d'abord que F'^{IJ} est équipotent à un élément de $\tilde{\mathcal{M}}_o$. En effet, soit $u = (C^{\bullet},\mu,\nu)$ un élément de S_o^{IJ} . Notons U_u l'ensemble des couples

$$(F,(t(i))_{i\in\alpha(F)_o}) \in N'\times \bigcup_{I^\bullet\in I} C^{I_o^\bullet} = U_C$$

tels que $\mu(F) = (F,t,\hat{e})$ soit défini, et U_ν l'ensemble des couples

$$(F',(t'(i))_{i\in\alpha(F')_o}) \in N'\times \bigcup_{I^\bullet\in J} C^{I_o^\bullet} = U'_C.$$

tels que $(\hat{e}',t',F') = \nu(F')$ soit défini. La surjection γ: $u \to (C^\bullet,U_\mu,U_\nu)$ est une bijection de S_o^{IJ} sur une partie de $U = N'_o\times \bigcup_{C\in M_o} (U_C\times U'_C)$. Comme $\hat{M}_o$ est un univers, les relations

$$N' \in \hat{M}_o,\quad I_o^\bullet \in M_o \text{ si } I^\bullet \in I\cup J\ ,\quad I\cup J\in\hat{M}_o,\quad M\in\hat{M}_o$$

entraînent

$$U_C \in \hat{M}_o\ ,\quad U'_C \in \hat{M}_o \quad\text{et}\quad U\in \hat{M}_o.$$

Par suite l'ensemble S_o^{IJ} est équipotent à un élément de $\hat{M}_o$. Le foncteur p''^{IJ} étant fidèle, S^{IJ} (et a fortiori F'^{IJ} et F^{IJ}) est équipotent à un élément de $\hat{M}_o$.

- Posons $p = (S^{IJ},\iota,F'^{IJ})$ et $P = (\hat{S}^{IJ},\iota,\hat{F}'^{IJ})$. D'après la proposition 5, P est à $\hat{M}_o$-produits, de sorte que P est aussi à S^{IJ}-produits. Par ailleurs p est à noyaux en vertu de la proposition 6; la démonstration de cette proposition signifie que tout noyau n appartient à X' et, X' définissant une sous-catégorie de $\hat{F}'^{IJ}$, il s'ensuit

$$X'.n \subset X'.X' \subset X'\ .$$

Enfin la proposition 8 affirme que $p'^{IJ} = p''^{IJ}.P$ est un foncteur $(M,X'.F_o'^{IJ})$-engendrant, et la proposition 7 que $P(X')\subset X'$ est formé de p''^{IJ}-monomorphismes. Des cas particuliers (§ 1) de la proposition 1, il résulte que p admet un adjoint, i.e. que S^{IJ} est à F'^{IJ}-projections. - On démontre les deux autres assertions d'une façon analogue, en prenant pour p respectivement le foncteur injection canonique de F^{IJ} vers S^{IJ} et de F^{IJ} vers F'^{IJ}. ∇

Soit $(\hat{u},\phi,u)$ un (F^{IJ},S^{IJ})-projecteur, où $u = (C^{\bullet},\mu,\nu)$. En général, ϕ n'est pas injectif. Mais, si u appartient à F'^{IJ}, on montre que ϕ est injectif, de sorte qu'il existe également un (F^{IJ},F'^{IJ})-projecteur

$$(u',\eta,u), \quad \text{où } u' = (C'^{\bullet},\mu',\nu') ,$$

tel que $C^{\bullet}$ soit une sous-catégorie de $C'^{\bullet}$ et que η soit le foncteur injection canonique de $C^{\bullet}$ vers $C'^{\bullet}$. Dans ce cas, on appelle u' une *(I,J)-complétion de* u . Supposons de plus que μ et ν soient les surjections vides; u' est aussi appelé *(I,J)-complétion libre de* $C^{\bullet}$. Cette définition est justifiée car, si q désigne le foncteur projection canonique de F^{IJ} vers F, on voit facilement que u' est une q-structure libre associée à $C^{\bullet}$.

REFERENCES.

1. Sur l'existence de structures libres et de foncteurs adjoints, *Cahiers de Topologie et Géométrie différentielle*, IX, 1 et 2, Dunod, Paris (1967), 33-186.

2. Esquisses et types de structures algébriques, *Bul. Inst. Polit. Iaşi*, XIV (XVIII), 1-2 (1968), 1-14.

3. *Catégories et structures*, Dunod, Paris (1965).

4. *Maîtrise de Mathématiques, C3, Algèbre*, Centre de Documentation Universitaire C.D.U.-S.E.D.E.S., Paris (1968).

5. Structures quasi-quotients, *Math. Ann.* 171 (1967), 293-363.

CATEGORIES OF GROUP EXTENSIONS

by

K. W. Gruenberg

§ 1

Our aim is to sketch the beginnings of a theory that studies a fixed group G in terms of all the extensions that can be constructed over G. We limit ourselves here to the extensions with abelian kernels. Even so, we uncover connexions between known but apparently unrelated aspects of group cohomology and group theory proper. The theory seems to provide the right kind of framework within which to seek extensions of these results and to point the way to new areas that may be worth exploring.

Details of everything that follows will be found in a set of lecture notes (probably to be called "cohomological topics in group theory") that I am preparing for this same series of Springer Lecture Notes.

§ 2

If G is a given group, let $(\underline{G})$ denote the category whose objects are all extensions

$$(A|E): \ 1 \longrightarrow A \longrightarrow E \longrightarrow G \longrightarrow 1 \ ,$$

with A abelian, and in which a morphism $(A|E) \longrightarrow (A_1|E_1)$

is a pair of group homomorphisms (α,σ), so that $\alpha: A \longrightarrow A_1$, $\sigma: E \longrightarrow E_1$ and the appropriate diagram is commutative. Of course, α is then necessarily a G-module homomorphism.

Let $co(A|E)$ denote the cohomology class in $H^2(G,A)$ determined by $(A|E)$. We can obtain this class as follows: Take any free presentation

$$\underline{(1)}. \quad 1 \longrightarrow R \longrightarrow F \xrightarrow{\pi} G \longrightarrow 1 \qquad (1)$$

(i.e., F is a free group) and lift π to $\theta: F \longrightarrow E$. Then θ_R (restriction to R) is a homomorphism: $R \longrightarrow A$ and this gives rise to $co(A|E)$. (Use MacLane's theorem that

$$Der(F,A) \longrightarrow Hom_F(R,A) \longrightarrow H^2(G,A) \longrightarrow 0$$

is exact.)

Remark. If F is free on a set in one-one correspondence with the non-unit elements of G, then θ_R is exactly a factor-set in the sense of Schreier.

If $(\alpha,\sigma): (A|E) \longrightarrow (A_1|E_1)$, then clearly $\alpha^*: co(A|E) \longmapsto co(A_1|E_1)$.

We now introduce a second category, $\mathcal{Q}_G$, whose objects are pairs (A,x), where $A \in Mod_G$ (= right G-modules) and $x \in H^2(G,A)$. A morphism is a module homomorphism $\alpha: A \longrightarrow A_1$ so that $x\alpha^* = x_1$. ($\mathcal{Q}_G$ is the category of $H^2(G, \ )$-pointed objects of Mod_G in MacLane's terminology: [4], p.53.) Then

$$\Gamma : (A|E) \longrightarrow (A, co(A|E))$$

is a functor from $\binom{G}{-}$ to $\mathcal{Q}_G$.

(2). Γ is surjective (= full and representative).
This is a completely elementary result. (Cf.,e.g., [1], p.179.) Nevertheless, it leads to very rapid proofs of some important results in group theory: e.g., the splitting theorem of Schur, the centre-commutator theorem of Schur, the Magnus representation theorem for F/[R,R].

The surjectivity of Γ is also constantly needed in what follows. In particular, for the characterization of epimorphisms and monomorphisms:

(3). (α,σ) is an epimorphism in $\binom{G}{-}$ if, and only if, α is an epimorphism in $\mathcal{Q}_G$ if, and only if, α is an epimorphism in Mod_G. Ditto for monomorphisms.

We may now define projectives and injectives in the usual manner.

(4). $(A|E)$ is injective in $\binom{G}{-}$ if, and only if, $\Gamma(A|E)$ is injective in $\mathcal{Q}_G$ if, and only if, A is injective in Mod_G.

Thus both our categories have enough injectives. But the injectives do not seem nearly as interesting as the projectives and we shall say nothing more about injectives here.

It is easy to manufacture examples of projectives.

Take any free presentation, as in (1) above, and write $\bar{R} = R/[R,R]$, $\bar{F} = F/[R,R]$. Then $(\bar{R}|\bar{F})$ is projective in $(\frac{G}{-})$ and $(\bar{R},\chi)$, where $\chi = \mathrm{co}(\bar{R}|\bar{F})$, is projective in $\mathcal{Q}_G$. In fact, these objects can be regarded as "free" in a certain natural sense. In any case, both categories have enough projectives.

It is difficult to say anything non-trivial about the relation between different free objects. The following is a simple consequence of Schanuel's lemma:

(5). *If $(\bar{R}_1|\bar{F}_1)$, $(\bar{R}_2|\bar{F}_2)$ are free objects, then*

$$\bar{R}_1 \oplus V_2 \simeq \bar{R}_2 \oplus V_1 \quad ,$$

where V_i is G-free of rank = rank F_i.

A more profitable project turns out to be the comparison of *minimal projectives*: The projective $(A|E)$ is called minimal if every epimorphism of $(A|E)$ to a projective is necessarily an isomorphism (and $A \neq 0$).

We first state a characterization of projectives:

(6). *The following statements are equivalent:*

(i) $(A|E)$ is projective in $(\frac{G}{-})$;

(ii) $\Gamma(A|E) = (A,x)$ is projective in $\mathcal{Q}_G$;

(iii) there exists a free pair $(\bar{R},\chi)$ so that

$(\bar{R},\chi) = (A,x) \oplus (P,0)$. ($\mathcal{Q}_G$ has products!)

Moreover, the module P in (iii) is necessarily G-projective.

It follows that a projective (A,x) is minimal if, and only if, there does not exist a splitting of the form

$$(A,x) = (B,y) \oplus (P,0) \quad .$$

If G is finite, $H^2(G, \text{projective}) = 0$ and so we have:

(7). For finite G, the projective (A,x) is minimal if, and only if, A has no projective direct summand.

It is often necessary to restrict the discussion to certain subcategories of $\mathcal{Q}_G$ corresponding to a change of scalars. Let K be a commutative ring and $\mathcal{Q}_{KG}$ the category of all pairs (A,x), with $A \in \text{Mod}_{KG}$ and morphisms to be KG-homomorphisms. A similar definition for $(\frac{KG}{\ })$.

If $A \in \text{Mod}_G$, let $A_{(K)} = A \underset{\mathbb{Z}}{\otimes} K$ and $A \longrightarrow A_{(K)}$ give $x \longmapsto x_{(K)}$. Then $(A,x) \longrightarrow (A_{(K)}, x_{(K)}) = (A,x)_{(K)}$ is a functor: $\mathcal{Q}_G \longrightarrow \mathcal{Q}_{KG}$.

Everything we have done so far works, with suitable modifications, in $\mathcal{Q}_{KG}$ and $(\frac{KG}{\ })$. In particular, the "K-free objects" in $\mathcal{Q}_{KG}$ are of the form $(\bar{R},\chi)_{(K)}$.

§ 3

Henceforth we shall assume G is finite.

Our concern will be with following coefficient rings: fields, $\mathbb{Z}$, $\mathbb{Z}_{(p)}$ = local ring at p, $\mathbb{Z}_p$ = p-adic integers.

Suppose K is a field of characteristic dividing $|G|$ and (A,x), (B,y) are finitely generated minimal projectives. If $(\bar{R},\chi)_{(K)}$, $(\bar{S},\tau)_{(K)}$ are K-free pairs with epimorphisms to (A,x), (B,y), respectively, then by (5) and (6) (iii),

$$A \oplus P \oplus (KG)^m \simeq B \oplus P' \oplus (KG)^n ,$$

where P, P′ are KG-projective. Since A,B have no KG-projective summands (by (7)), the Krull-Schmidt theorem implies

<u>(8). A ≃ B.</u>

(If the characteristic of K is prime to $|G|$, then (A,x) is minimal projective if, and only if, x = 0 and A is inducible.)

<u>(9). If K is a field of characteristic dividing $|G|$, then any two finitely generated minimal projectives are isomorphic.</u>

To complete the proof of (9), we need more cohomological information about A and B.

In view of (6)(iii), the cohomology in A is the same as that in $\bar{R}_{(K)}$. Now we have the following result.

<u>(10). Let K be any commutative ring. Then</u>

<u>(i) $H^{q+2}(G,\bar{R}_{(K)}) \overset{d^2}{\underset{\sim}{\longleftarrow}} H^q(G,K)$, for all $q > 0$;</u>

<u>(ii) $H^2(G,\bar{R}_{(K)}) \overset{\sim}{\longrightarrow} K/nK$, where $n = |G|$, and</u>

<u>$\chi_{(K)} \longmapsto 1 + nK$;</u>

<u>(iii) $H^1(G,\bar{R}_{(K)}) \overset{\bar{d}}{\longleftarrow} g^G$, where g is the augmentation ideal of KG.</u>

<u>Remark.</u> If we had used Tate cohomology, then (ii) and (iii) would be special cases of (i). The case $K = \mathbb{Z}$ is due to Tate (cf. Kawada [3]).

<u>Proof.</u> The homomorphism $\pi: F \longrightarrow G$ yields $0 \longrightarrow r \longrightarrow KF \longrightarrow KG \longrightarrow 0$ and so, if f, g are the augmentation ideals of KF, KG, respectively,

(11). $0 \longrightarrow r/fr \longrightarrow f/fr \longrightarrow g \longrightarrow 0$ is exact in Mod_{KG}. We know f/fr is KG-free on all $(1 - x_i) + fr$, where (x_i) freely generate F and that $r/fr \cong \bar{R}_{(K)}$. We also have the exact sequence

(12). $0 \longrightarrow g \longrightarrow KG \longrightarrow K \longrightarrow 0$. Obviously (11) and (12) yield (i); also

$$(KG)^G \longrightarrow K \longrightarrow H^1(g) \overset{\sim}{\longrightarrow} H^2(\bar{R}_{(K)})$$

gives (ii) since $(KG)^G = K\tau$, where $\tau = \sum_{x \in G} x$. Finally, (11) yields (iii) since $(\mathfrak{f}/\mathfrak{f}\mathfrak{r})^G \longrightarrow \mathfrak{g}^G$ in the zero map.

We may now complete (9). So K is a field of characteristic dividing $|G|$. Thus $nK = 0$ and so x is a basis of $H^2(G,A)$ and y a basis of $H^2(G,B)$. Let ϕ be the isomorphism (8). Then

$$x\phi^* = ky$$

and hence $\frac{1}{k}\phi$ gives an isomorphism: $(A,x) \xrightarrow{\sim} (B,y)$.

Remark. The proof of (9) obviously also works if K is $\mathbb{Z}_p$, for any p.

§ 4

Result (10) can be used to give a purely cohomological characterization of projectives in $\mathcal{Q}_{KG}$, where K is a field.

We first show that every element (A,x) in $\mathcal{Q}_{KG}$ determines in a natural way an element in $H^1(G,A)$.

Given $(A|E)$, let $T = (t_i)$ be a transversal (= set of representatives of the cosets) of A in E and for each e in E, let

$$t_i e = a_{i,e}\, t_{i(e)} \quad .$$

Put

$$ed_T = \prod_i t_{i(e)}^{-1}\, a_{i,e}\, t_{i(e)} \quad .$$

Then d_T is a derivation of E in A. If S is second

transversal, then d_S is cohomologous to d_T. So we obtain an element, call it the transfer class, in $H^1(E,A)$. (Note that when A is central in E, then d_T is the ordinary transfer homomorphism.) In our case of K a field with char K dividing $|G|$, $Ad_T = 1$ and so the transfer class can really be thought of as an element in $H^1(G,A)$.

Remark due to B. Wehrfritz: When $A \xrightarrow{|G|} A$ is an automorphism, Ker d_T is a subgroup complementary to A and since all such subgroups have this form, they are all conjugate: another proof of Schur's theorem!

(13). If $\tau = \sum_{x \in G} x \in g^G$, then the isomorphism (10)(iii) maps τ to the transfer class of $(\bar{R},\chi)_{(K)}$.

(14). If the field K has characteristic dividing $|G|$, then (A,x) is projective in $\mathcal{Q}_{KG}$ if, and only if,

(i) $H^1(G,A)$ has dimension one over K and the transfer class $\neq 0$;

(ii) $H^2(G,A)$ has dimension one over K and $x \neq 0$.

Proof. One direction is a trivial consequence of (10) and (6). Conversely, assume (i) and (ii) and pick an epimorphism ϕ: $(\bar{R},\chi)_{(K)} \longrightarrow (A,x)$.

Then ϕ induces isomorphisms $H^i(\bar{R},_{(K)}) \xrightarrow{\sim} H^i(A)$, $i = 1,2$, and hence $H^2(P) = 0$ where $P = \mathrm{Ker}\phi$. Thus P is KG-projective (cf. [5], chaper 9) and hence $\bar{R}_{(K)}$ splits over P (K is a field!).

Somewhat similar arguments yield

(15). (A,x) is projective in $\mathcal{Q}_G$ if, and only if, for all p and any Sylow p-subgroup G_p,

(i) $H^1(G_p,A) = 0$,

(ii) $H^2(G_p,A)$ is cyclic on $x\mathrm{Res}$, of order $|G_p|$; and

(iii) A is $\mathbb{Z}$-free.

There is also a localization principle:

(16). If (A,x) is finitely generated, then (A,x) is projective in $\mathcal{Q}_G$ if, and only if, $(A,x)_{(\mathbb{Z}_p)}$ is projective in $\mathcal{Q}_{\mathbb{Z}_p G}$, for all p.

§ 5

We return now to minimal projectives. Let $\mathbb{Z}_G = \bigcap_{p/|G|} \mathbb{Z}_{(p)}$ and write $M_G = M \otimes \mathbb{Z}_G$. We shall say (A,x), (B,y) are in the same genus if $(A,x)_G \simeq (B,y)_G$.

(17). Any two finitely generated minimal projectives are in the same genus.

We omit the proof here, but I wish to acknowledge having had crucial help from Irving Reiner with one part of this proof.

Finally, we mention a surprising connexion of some of this theory with much earlier work of Gaschütz.

Let K be a field of characteristic dividing $|G|$. We shall call $(A|E)$ a <u>Frattini extension</u> if $A \leq Fr(E)$, the Frattini group of E. The Frattini $(A|E)$ is maximal if any epimorphism from another Frattini to $(A|E)$ is necessarily an isomorphism. It is not at all obvious that maximal Frattini extensions even exist. In fact they do and any two are isomorphic. This was established by Gaschütz (for $K = \mathbb{F}_p$) in 1954 [2]. It follows from our theory as a consequence of (9) and the following theorem:

<u>(18)</u>. If K is a field of characteristic dividing $|G|$, then $(A|E)$ is maximal Frattini if, and only if, $(A|E)$ is minimal projective.

REFERENCES

[1] Artin, E. and J. Tate, *Class Field Theory*, Benjamin, New York, 1967.

[2] Gaschütz, W., "Über modulare Darstellungen endlicher Gruppen, die von freien Gruppen induziert werden", *Math. Z.* 60; 274-286. (1954).

[3] Kawada, Y., "Cohomology of group extensions", *J. Fac. Sc. Univ. Tokyo*, 9; 417-431. (1963).

[4] MacLane, S., "Categorical Algebra", *Bull. Amer. Math. Soc.*, 71; 40-106. (1965).

[5] Serre, J.-P., *Corps locaux*, Hermann, Paris, 1962.

ALGEBRAS GRADED BY A GROUP

by

Max A. Knus*

INTRODUCTION

A Brauer theory for $\mathbb{Z}/2\mathbb{Z}$-graded algebras was developed by C. T. Wall in [7], (See also H. Bass, Lectures on algebraic K-theory, Bombay, 1967, [3]). In this paper, we try to define such a theory for algebras graded by an arbitrary abelian group G. We first define graded central simple algebras and give some examples. This class of algebras is closed under a suitable graded tensor product. Before defining the Brauer group, we prove a structure theorem for graded central simple algebras. Finally, we study a class of graded algebras which generalize Clifford algebras.

My sincere thanks are due to Michel André for many suggestions and discussions.

---

*This work was carried out under a grant from the National Science Foundation and a fellowship from the Schweizerische Nationalfonds.

1. GRADED CENTRAL SIMPLE ALGEBRAS

By algebra, we shall mean a finite dimensional associative algebra A with unit over a commutative field K.

Let G be a group, written additively, but not necessarily abelian. A G-graded algebra A is an algebra which is given together with a direct sum decomposition as a module

$$A = \bigoplus_{g \in G} A_g$$

where the A_g's are subspaces of A, in such a way that

$$A_g A_h \subset A_{g+h} \quad .$$

The elements of K are homogeneous of degree zero. A homomorphism of graded algebras is a homomorphism of algebras $\varphi: A \to B$ such that $\varphi(A_g) \subset B_g$, $g \in G$. A subspace I of A is graded if it is the direct sum of the intersections $I \cap A_g$.

We call the graded algebra A simple if there are no proper graded (two-sided) ideals. A graded simple algebra is not necessarily a simple algebra, but the following easy generalization of the Theorem of Maschke is true.

Theorem 1.1

Let A be a simple graded algebra. If the characteristic of K does not divide the dimension of A over K, then A is a semisimple algebra.

Proof. Let $x = \sum_{g \in G} x_g$, $x \neq 0$, be in the radical of A. Since A is simple graded, we may assume that $x_0 = 1$. Hence $\mathrm{Trace}(x) = \mathrm{Dim}_K A$ is not zero and x is not nilpotent. Therefore the radical of A must be zero.

Let $\varphi: G \times G \longrightarrow K^*$ be a pairing of G in the multiplicative group of K. This means

$\varphi(g_1 + g_2, h) = \varphi(g_1, h)\, \varphi(g_2, h)$ and

$\varphi(g, h_1 + h_2) = \varphi(g, h_1)\, \varphi(g, h_2)$ for g_i, h_i in G.

In particular $\varphi(0,h) = \varphi(g,0) = 1$.

Let $A = \bigoplus_{g \in G} A_g$ be a G-graded algebra. In order to simplify the notation, we shall write $\varphi(a,b)$ for $\varphi(\text{degree } a, \text{degree } b)$.

Let now φ be a fixed pairing. We say that the graded algebra A is central if the only homogeneous elements x of A such that, $xa = \varphi(x,a)ax$ for all homogeneous a in A or $ax = \varphi(a,x)xa$ for all homogeneous a in A are in K.

Examples of graded central simple algebras

1.2 Let $H^2(G,K^*)$ be the second cohomology group of G with coefficients in K^*, where G operates trivially on K^*. Take a normalized cocycle f in any class of $H^2(G,K^*)$. We define an associative multiplication on the space of all formal linear combinations

$$\sum_{g\in G} \alpha_g x_g$$

with coefficients $\alpha_g \in K$ by setting

$$\Big(\sum_{g\in G} \alpha_g x_g\Big)\Big(\sum_{h\in G} \beta_h x_h\Big) = \sum_{g,h\in G} \alpha_g \beta_h f(g,h) x_{g+h} .$$

We shall denote the G-graded algebra thus constructed by K_fG. If f is the trivial cocycle, K_fG is simply the group algebra KG of G over K. We remark that two cohomologous cocycles define isomorphic graded algebras. The resulting class is called an algebra extension of G over K. See [8] for more details.

An algebra extension is obviously graded simple. Suppose now that G is abelian and that f is an abelian cocycle, i.e.

$$f(g,h) = f(h,g) \qquad g,h \in K .$$

The algebra extension is central if the given pairing φ satisfies the following condition:

Non-degeneracy condition: for every $g \in G$, $g \neq 0$, there exists $h \in G$ such that $\varphi(g,h) \neq 1$.

In particular, the group algebra then is graded central simple.

It is easy to construct a nondegenerate pairing if the group G is finite abelian and K is G-cyclic (this means that, if G has an element of order n, then K contains a primitive n-th root of 1). Let

$$(*) \qquad G = \oplus \mathbb{Z}/p_i^{r_i}\mathbb{Z}$$

be the direct sum decomposition of G such that $p_1 \leq p_2 \leq \dots$, p_i primes and $r_m \geq r_{m+1} \geq \dots \geq r_q$ if $p_m = p_{m+1} = \dots = p_q$. Choose a primitive $p_i^{r_i}$-th root ω_i for all i appearing in the direct sum (*). For $g,h \in G$ define

$$\omega^g = \prod_i \omega_i^{g_i} \quad \text{and} \quad \omega^{gh} = \prod_i \omega_i^{g_i h_i}$$

where $g_i \in \mathbb{Z}$ is a representative of the class of the i-th component of g in the decomposition (*). The map $\varphi\colon G \times G \longrightarrow K^*$ defined by $\varphi(g,h) = \omega^{gh}$ is a pairing satisfying the non-degeneracy condition. In the first version of this paper, we considered only this special pairing. Michel André suggested the use of an abstract pairing.

1.3. Let V be a finite dimensional G-graded vector space over K, G any abelian group. The algebra $\mathrm{End}_K(V)$ is simple, graded, therefore graded simple. Furthermore it is not difficult to see that $\mathrm{End}_K(V)$ is graded central (for any pairing φ).

Remark. If G is a finite cyclic group and K is algebraically closed and G-cyclic, then one can easily show that a graded algebra A which is central simple (non-graded) is isomorphic as a graded algebra to $\mathrm{End}_K(V)$ for some G-graded vector space V. One proceeds as follows. Let ω be a primitive n-th root of 1, where n is the order of G. The map $\psi: A \longrightarrow A$ defined by

$$\psi: a_g \longrightarrow \omega^g a_g \quad , \quad a_g \in A_g$$

is a non-trivial automorphism of A (if A is non-trivially graded ! The result is classical if the grading of A is trivial). The algebra is central simple, therefore ψ is inner,

$$\psi a_g = u a_g u^{-1}$$

for some $u \in A$. The element u is determined up to a scalar multiple. By definition of ψ, the graded structure of A is completely determined by u,

$$A_g = \{a \in A \mid \omega^g au = ua\} \quad .$$

In particular, we see that u belongs to A_0. The automorphism ψ^n is trivial, therefore u^n is central and thus is equal to an element $\alpha \neq 0$ of K. Since u is only determined up to a scalar multiple, we may assume that $u^n = 1$. The field K is algebraically closed, hence we can identify A with the endomorphism algebra $\mathrm{End}_K(V)$ of a vector space

V over K. The minimal polymonial $Q_u(t)$ of the endomorphism u is clearly equal to $t^n - 1$. The roots of $Q_u(t)$ are ω^g, $g \in G$. Set

$$V_g = \{x \in V \mid u(x) = \omega^g x\} \quad .$$

Since $Q_u(t)$ is a product of distinct linear factors over K,

$$V = \bigoplus_{g \in G} V_g \quad .$$

Finally one verifies that

$$A_g = \{\alpha \in \mathrm{End}_K(V) \mid \alpha: V_h \longrightarrow V_{g+h},\ \forall\ h \in G\}$$

as it should be.

2. GRADED TENSOR PRODUCT

A graded tensor product $A \otimes B$ of two G-graded algebras A and B is the usual graded vector space

$$A \otimes B = \bigoplus_{g \in G} (A \otimes B)_g$$

where

$$(A \otimes B)_g = \bigoplus_{h+h'=g} A_h \otimes {}_K B_{h'}$$

with the multiplication

$$(a \otimes b)\ (a' \otimes b') = \varphi(b,a')aa' \otimes bb'$$

$$a,a' \in A \quad \text{and} \quad b,b' \in B \quad .$$

The map φ is a pairing of G in K^*. The product thus defined is associative.

Example. Let $G = \mathbb{Z}/2\mathbb{Z}$. The usual graded tensor product corresponds to $\varphi(g,h) = (-1)^{gh}$.

Let φ now be a fixed pairing. All tensor products will be taken with respect to φ. Let A, B and C be graded algebras. The tensor product is associative,

$$(A \otimes B) \otimes C \cong A \otimes (B \otimes C) \quad ,$$

but in general not commutative, $A \otimes B \ncong B \otimes A$.

Let V and V′ be finite dimensional G-graded vector spaces over K. The algebra $End_K(V)$ is graded and

$$End_K(V) \otimes End_K(V') \cong End_K(V \otimes V')$$

if we define

$$(f \otimes f')(x \otimes x') = \varphi(f',x) fx \otimes f'x' \quad .$$

Proposition 2.1

If A and B are graded central simple algebras, then $A \otimes B$ is also graded central simple.

Proof. The proof for the non-graded case, given for example in [4], can be used with some trivial modifications.

Let A be a graded algebra. The opposite algebra A^* of A is identic to A as a vector space and has the multiplication

$$a^*b^* = \varphi(a,b)ba \quad .$$

The algebra A^* is obviously central simple if A is so.

There is a natural homomorphism of graded algebras

$$\mu : A \otimes A^* \longrightarrow \mathrm{End}_K(A)$$

defined by $(a \otimes b^*)x = \varphi(b^*,x)axb$ if G is abelian.

Proposition 2.2

The natural map μ is an isomorphism if A is central simple.

Proof. The same as in the non-graded case. See for example [4].

3. A STRUCTURE THEOREM FOR G-GRADED ALGEBRAS

A pairing $\varphi: G \times G \longrightarrow K^*$ is called symmetric if $\varphi(g,h) = \varphi(h,g)$ for all g,h in G. Suppose that G is abelian.

Theorem 3.1

Let φ be a symmetric pairing of G satisfying the non-degeneracy condition for any finite subgroup of G. Then a graded central simple algebra A such that $\mathrm{char}\, K \nmid \dim_K A$ is isomorphic to a graded algebra of the type

$$K_f(H) \otimes \mathbb{M}_n(D,H') \quad .$$

H and H' are subgroups of G such that $G \supseteq H \times H'$. f is an abelian class in $H^2(H,K^*)$ and $K_f(H)$ is the corresponding extension of H over K. $\mathbb{M}_n(D,H')$ is a matrix algebra over a division ring D, graded by H'.

<u>Proof</u>. Let Z be the ungraded center of A,

$$Z = \{x \in A |\ xa = ax \text{ for all } a \in A\} \ .$$

The space Z is certainly graded, $Z = \bigoplus_{h\in H} Z_h$, where $H = \{h \in G |\ Z_h \neq 0\}$. The product of two non-zero homogeneous elements of Z is not zero, since A is graded simple. Therefore H is a subgroup of G. The subgroup H must be finite because A is finite dimensional. The space Z_0 contains K and is contained in the graded center of A, hence $Z_0 = K$. Let $x_h \neq 0$ be in Z_h, $h \in H$. If n is the order of h in G, x_h^n is different from zero and is contained in Z_0, therefore $x_h^n = \alpha$, $\alpha \in K$, $\alpha \neq 0$. We have $Z_h = Z_h x_h^n \subset Z_0 x_h = Kx_h$, hence all Z_h, $h \in H$, are one dimensional. Choose $x_h \neq 0$ in each Z_h. We know that $x_h x_{h'} \neq 0$, therefore the equality

$$x_h x_{h'} = f(h,h') x_{h+h'}$$

defines an abelian 2-cocycle f of H with value in K^*. Another choice of the x_h's would define a cocycle in the same cohomology class. We have thus constructed the part $K_f(H)$.

Let now B be the graded centralizer of Z in A,

$$B = \{x \in A |\ xz = \varphi(x,z) zx \ \ \forall z \in Z \text{ or}$$
$$zx = \varphi(z,x) xz, \ \forall z \in Z\} \ .$$

As $zx = xz$ for all z in Z,

$$B = \{x \in A \mid \varphi(z,x) = 1 \quad \text{for all} \quad z \in Z\}$$

(φ is symmetric !). We show that $B \cong \mathrm{Hom}_{Z\otimes Z^*}(Z,A)$. An element $f \in \mathrm{Hom}_{Z\otimes Z^*}(Z,A)$ is a map $f\colon Z \longrightarrow A$ such

$$f((z_1 \otimes z_2^*)x) = \varphi(f,z_1 \otimes z_2^*)z_1 \otimes z_2^* f(x)$$

$$= \varphi(f,z_1 \otimes z_2^*)\varphi(z_2,fx)z_1 f(x) z_2 \quad .$$

Let $\rho\colon \mathrm{Hom}_{Z\otimes Z^*}(Z,A) \longrightarrow A$ be the map defined by $f \longrightarrow f(1)$. It follows from $f((z \otimes 1)1) = f((1 \otimes z)1)$ that $zf(1) = \varphi(z,f(1))f(1)z$ for all z in Z, hence $f(1)$ is in B. On the other side, the map $f_b\colon Z \longrightarrow A$ defined by $f(z) = zb$, $b \in B$, is a $Z \otimes Z^*$-homomorphism.

By hypothesis, the pairing φ satisfies the non-degeneracy condition on H. Therefore $Z = K_f(H)$ is graded central simple and, by <u>Proposition 2.2</u>, $Z \otimes Z^* = \mathrm{End}_K(Z)$. Hence $B \cong \mathrm{Hom}_{\mathrm{End}_K(Z)}(Z,A)$. By <u>Proposition A6</u> of [2],

the map

$$\sigma\colon Z \otimes_K \mathrm{Hom}_{\mathrm{End}_K(Z)}(Z,A) \longrightarrow A$$

given by $\sigma(x \otimes f) = f(x)$ is an isomorphism of vector spaces over K. The corresponding isomorphism $Z \otimes_K B \longrightarrow A$ is defined by $x \otimes f(1) = (x \otimes 1)f(1) = xf(1)$ (see [2′] <u>Theorem 3.1</u>.) which is an isomorphism of algebras. Hence we can identify $Z \otimes B$, ZB and A.

Let now x be in the graded center of B, $xb = \varphi(x,b)bx$ or $bx = \varphi(b,x)xb$ for $b \in B$. Since $\varphi(x,z) = 1$ and $xz = zx$ for $z \in Z$, we can write $xzb = \varphi(x,zb)zbx$ or $zbx = \varphi(zb,x)xzb$, therefore x is in K, the graded center of A, B is graded central. B is also ungraded central, because any element of the center of B would be in Z, and B and Z are disjoint over K. It is not difficult to see that B is graded simple. Since char $K \nmid \dim_K A$, char $K \nmid \dim_K B$ and by <u>Proposition 1.1.</u>, B is semisimple. A semisimple central algebra is simple, hence B is (ungraded) simple.

Let H' be the set of degrees of B, $H' = \{g \in G \mid \varphi(g,h) = 1, \forall\, h \in H\}$. H' is a subgroup of G. Suppose there exists $h \in H \cap H'$, $h \neq 0$. By definition of B, Z_h would be contained in B_h, but we know that Z and B are disjoint over K. Hence $H \cap H' = 0$. Remark that, if $Ag \neq 0$ for all $g \in G$, then $G = H \times H'$.

Suppose that K is algebraically closed and G-cyclic. The construction of <u>Example 1.1</u> gives a pairing satisfying the nondegeneracy condition for H. Furthermore $H^2(H,K^*) = 1$ if K is algebraically closed. If G is cyclic, we showed in <u>Example 1.2</u> that $M_n(K,H')$ is the endomorphism algebra of a vector space graded by H'.

Corollary 3.2

Let K be algebraically closed and G cyclic of prime order. Suppose that K is G-cyclic. A G-graded central simple algebra A such that $\operatorname{char} K \nmid \dim_K A$ is isomorphic to an algebra of the following types:

1) $\operatorname{End}_K(V)$ where V is graded by G
2) $KG \otimes \operatorname{End}_K(V)$ where $\operatorname{End}_K(V)$ is ungraded.

The proof follows from the remarks immediately above.

Remark. The hypothesis that $\operatorname{char} K \nmid \dim_K A$ is not necessary in Corollary 3.2. A direct proof can be given, following the proof of Lemma 4 in [7].

4. THE BRAUER GROUP

Let $\varphi: G \times G \longrightarrow K^*$ be a fixed pairing of abelian group G in a commutative field K. We say that two G-graded central simple algebras A and B are similar if there exist two G-graded vector spaces V_1 and V_2 such that

$$A \otimes \operatorname{End}_K(V_1) \cong B \otimes \operatorname{End}_K(V_2)$$

as graded algebras. Let $B(K,G)$ be the set of equivalence classes. By Proposition 2.1 and 2.2, the tensor product induces a group structure on $B(K,G)$. The classical (nongraded) Brauer group $B(K)$ is certainly contained in $B(K,G)$.

Example 4.1. Let K be algebraically closed and G cyclic of prime order. It follows from Corollary 3.2 that $B(K,G)$ is cyclic of order two. If G is any finite cyclic group and $\text{char } K = 0$, one can see that $B(K,G)$ contains a product of n copies of $\mathbb{Z}/2\mathbb{Z}$, where n is the number of primes appearing in the order of G.

5. A GENERALIZATION OF CLIFFORD ALGEBRAS

Let G be a finite abelian group and K a G-cyclic field. We shall use the pairing φ constructed in Example 1.2. Let K_nG be the n-fold graded tensor product (with respect to φ) of the group algebra KG with itself. If G is cyclic of order m, K_nG is described by generators e_i, $i = 1, 2, \ldots n$ subject to the relations

$$e_i^m = 1 \quad i = 1, 2, \ldots, n$$

$$e_ie_j = \omega e_je_i \quad i < j,$$

ω is a primitive m-th root of 1. The (non-graded) structure of K_nG, G cyclic, was described by Morris [6], when m is odd or m is even and ω has a square root in K. We shall give a complete description for arbitrary finite abelian groups.

Let KG^* be the opposite algebra of KG (remark that KG^* is the algebra extension of G over K corresponding to the cocycle φ). Write K_nG^* for the n-fold graded tensor product $KG^* \otimes \ldots \otimes KG^*$. The algebras K_nG and K_qG^* are connected by the following relations:

$$K_nG^* \ |\otimes|\ K_2G \cong K_{n+2}G$$

$$K_nG \ |\otimes|\ K_2G^* \cong K_{n+2}G^* \quad .$$

The sign $|\otimes|$ means ungraded (without φ !) tensor products. The proof [5] is similar to the proof given for Clifford algebras [1]. The algebras K_nG and K_qG^* are the Clifford algebras C'_n and C_q if G is the cyclic group of order two.

To describe completely the algebras K_nG and K_qG^*, it is therefore sufficient to know K_iG and K_iG^* for $i = 1,2$. Using the results of Morris [6], one obtains

$$K_1G = K \times \ldots \times K \quad n \text{ copies}$$

$$K_2G = M_n(K)$$

n is the order of G. If n is odd, it is easy to see that $K_iG = K_iG^*$. A more interesting case is when G is cyclic of order a power of 2, $n = 2^r$. We know that K contains a primitive 2^r-th root of $1, \omega$. Suppose ω does not have a square root in K. Then

$$K_1G^* = K(\xi) \times \ldots \times K(\xi) \qquad 2^{r-1} \text{ copies}$$

where $K(\xi)$ is the quadratic extension of K such that $\xi^2 = \omega$.

$$K_2G^* = \mathbb{M}_{2^{r-1}}(K_{\mathbb{H}})$$

where $K_{\mathbb{H}}$ is the quaternionic algebra over K given by generators ξ and η subject to the relations

$$\xi^2 = \eta^2 = \omega \quad \text{and}$$

$$\xi\eta = -\eta\xi \quad .$$

If ω has a square root in K, then $K_i G = K_i G^*$. Complete results and proofs are given in [5].

REF.

[1] Atiyah, M., Bott, R., and Shapiro, A., "Clifford modules", *Topology*, 3, (Supplement 1); 3-38, (1965).

[2] Auslander, M., and Goldman, O., "Maximal orders", *Trans. Amer. Math. Soc.*, 97; 1-24. (1960).

[2'] Auslander, M., and Goldman, O., "The Brauer group of a commutative ring", *Trans. Amer. Math. Soc.*, 97; 367-409. (1960).

[3] Bass, H., *Lectures on algebraic K-theory*, Bombay, (1967).

[4] Herstein, I., *Noncommutative rings*, Carus Publ. #15, MAA, Providence, (1968).

[5] Knus, M. A., *A generalization of Clifford algebras*, 10 p. (unpublished).

[6] Morris, A. M., "On a generalized Clifford algebra", *Quart. J. Math.*, (Oxford Ser.), 18; 7-12. (1967).

[7] Wall, C. T. C., "Graded Brauer groups", *J. Reine Angew. Math.*, 213; 187-199. (1964).

[8] Yamazaki, K., "On projective representations and ring extensions of finite groups", *J. Fac. Sci. Univ. Tokoyo, Sect. I*, 10; 147-195. (1964).

Diagonal Arguments and Cartesian Closed Categories

by

F. William Lawvere

The similarity between the famous arguments of Cantor, Russell, Gödel and Tarski. is well-known, and suggests that these arguments should all be special cases of a single theorem about a suitable kind of abstract structure. We offer here a fixed-point theorem in cartesian closed categories which seems to play this role. Cartesian closed categories seem also to serve as a common abstraction of type theory and propositional logic, but the author's discussion at the Seattle conference of the development of that observation will be in part described elsewhere ["Adjointness in Foundations", to appear in Dialectica, and "Equality in Hyperdoctrines and the Comprehension Schema as an Adjoint Functor", to appear in the Proceedings of the AMS Symposium on Applications of Category theory].

1. By a cartesian closed category is meant a category C equipped with the following three kinds of right adjoints: a right adjoint 1 to the unique

$$C \to \mathbf{1},$$

a right adjoint $\times$ to the diagonal functor

$$C \to C\times C,$$

and for each object A in C, a right adjoint $(\ )^A$ to the functor

$$C \xrightarrow{A\times(\ )} C.$$

The adjunction transformations for these adjoint situations, also assumed given, will be denoted by δ,π in the case of products and by λ_A,ϵ_A in the case of exponentiation by A. Thus for each X one has

$$X \xrightarrow{X\lambda_A} (A\times X)^A$$

and for each Y one has

$$A\times Y^A \xrightarrow{Y\epsilon_A} Y.$$

Given $f:A\times X \to Y$, the composite morphism

$$X \xrightarrow{X\lambda_A} (A\times X)^A \xrightarrow{f^A} Y^A$$

will be called the "λ-transform" of the morphism f. A morphism $h: X \to Y^A$ is the λ-transform of f iff the diagram

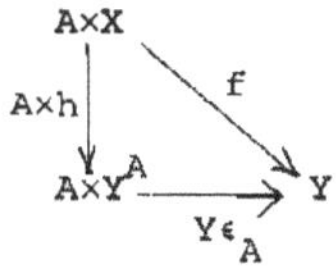

is commutative, showing in particular that f can be uniquely recovered from its λ-transform. Taking the case $X = 1$, one has that every $f: A \to Y$ gives rise to a unique $\ulcorner f \urcorner : 1 \to Y^A$ and that every $1 \to Y^A$ is of that form for a unique f. Since for every $a: 1 \to A$ one has (dropping the indices A,Y on ϵ when they are clear)

$$\langle a, \ulcorner f \urcorner \rangle \epsilon = a.f,$$

one calls ϵ the "evaluation" natural transformation; note however that we do not assume in general that f is determined by the knowledge of all its "values" a.f.

Although we do not make use of it in this paper, the usefulness of cartesian closed categories as algebraic versions of type theory can be further illustrated by assuming that the coproduct

$$2 = 1+1$$

also exists in C. It then follows (using the closed structure), that for every object A

$$A\times 2 = A+A$$

and so in particular that 2 is Boolean-algebra-object in C, i.e. that among the morphisms

$$2\times 2\times \ldots \times 2 \to 2$$

in C there are well determined morphisms corresponding to all the finitary (two-valued) truth tables, and that these satisfy all the commutative diagrams expressing the axioms of Boolean algebra. Equivalently, for each X the set

$$P_C(X) = C(X,2)$$

of "C-attributes of type X" becomes canonically an actual Boolean algebra, and varying X along any morphism of C induces contravariantly a Boolean homomorphism of attribute algebras. The morphisms $1 \to 2$ form $P_C(1)$ the Boolean algebra of "truth-values"; among these are the two coproduct injections which play the roles of "true" and "false". For any "constant of type X" $x:1 \to X$ and any attribute φ of type X, $x.\varphi$ is then a truth-value. Now noting that

$$X \times 2^X \xrightarrow[(2)\epsilon_X]{} 2$$

is a "binary operation" we could write it between its arguments, so that we have

$$x \epsilon \ulcorner\varphi\urcorner = x.\varphi,$$

an equality of truth values; thus if we think of $\ulcorner\varphi\urcorner:1 \to 2^X$ as the constant naming the subset of X corresponding to the attribute φ, one sees that the above equation expresses the usual "comprehension" axiom.

Returning to our immediate concern, we define a morphism $g:X \to Z$ to be <u>point-surjective</u> iff for every $z:1 \to Z$ there exists $x:1 \to X$ with $xg = z$. This does not imply that g is necessarily "onto the whole of Z", since there may be few morphisms with domain 1; for example if (as in the next section) X and Z are set-valued functors, then a natural transformation g is point-surjective if every element of the <u>inverse limit</u> of Z comes from an element of the inverse limit of X. In case Z is of the form Y^A, an even weaker notion of surjectivity can be considered, which in fact suffices for our fixed point theorem. Namely

$$X \xrightarrow{g} Y^A$$

will be called <u>weakly point-surjective</u> iff for every $f:A \to Y$ there is x such that for every $a:1 \to A$

$$\langle a, xg \rangle \epsilon = a.f$$

Finally we say that an object Y has the <u>fixed-point property</u> iff for every endomorphism $t:Y \to Y$ there is $y:1 \to Y$ with $y.t = y$.

<u>Theorem</u> In any cartesian closed category, if there exists an object A and a weakly point-surjective morphism

$$A \xrightarrow{g} Y^A$$

then Y has the fixed point property.

Proof: Let $\bar{g}$ be the morphism whose λ-transform is g. Then for any $f:A \to Y$ there is $x:1 \to A$ such that for all $a:1 \to A$

$$\langle a,x \rangle \bar{g} = a.f.$$

Now consider any endomorphism t of Y and let f be the composition

$$A \xrightarrow{A\delta} A\times A \xrightarrow{\bar{g}} Y \xrightarrow{t} Y;$$

thus there is x such that for all a

$$\langle a,x \rangle \bar{g} = \langle a,a \rangle \bar{g}t$$

since $a(A\delta) = \langle a,a \rangle$. But then $y = \langle x,x \rangle \bar{g}$ is clearly a fixed point for t.

The famed "diagonal argument" is of course just the contrapositive of our theorem. Cantor's theorem then follows with $Y = 2$.

Corollary If there exists $t:Y \to Y$ such that $yt \neq y$ for all $y:1 \to Y$ then for no A does there exist a point-surjective morphism

$$A \to Y^A$$

(or even a weakly point-surjective morphism).

2. Russells Paradox does not presuppose that set theory be formulated as a higher type theory; that is, for A the set-theoretical universe, we do not need 2^A for the argument. In fact we need only apply the proof of our theorem, with $\bar{g}:A\times A \to 2$ as the set-theoretical membership relation, dispensing with g entirely. That is, more generally, our theorem could have been stated and proved in any category with only finite products (no exponentiation) by simply phrasing the notion of (weak) point-surjectivity as a property of a morphism

$$A\times X \to Y;$$

however discovering the latter form (or at least calling it surjectivity!) seems to require thinking of such a morphism as a family of morphisms $A \to Y$ indexed by the elements of X, suggesting that a closed category is the "natural" setting for the theorem.

In fact the more general form of the theorem just alluded to (for categories with products) follows from the cartesian closed version which we have proved, by virtue of the following remark. Notice that it would suffice to assume C small (just take the full closure under finite products of the two objects A,Y)

<u>Remark</u> Any small category C can be fully and faithfully embedded in a cartesian closed category in a manner which preserves any products or exponentials which may exist in C.

Proof: We consider the usual embedding

$$C \subseteq \mathcal{S}^{C^{op}}$$

which identifies an object Y with the contravariant set-valued functor

$$X \rightsquigarrow C(X,Y).$$

By "Yoneda's Lemma" one has for any functor Y and any object A that the value at A of Y

$$AY \tilde{\to} \mathcal{S}^{C^{op}}(A,Y)$$

where the right hand side denotes the set of all natural transformations from (the functor corresponding to) A into Y, so that in particular the embedding is full and faithful. It is then also clear that the embedding preserves products (in particular if 1 exists in C it corresponds to the functor which is constantly the one-element set, which is the 1 of $\mathcal{S}^{C^{op}}$). For any two functors A,Y the functor

$$C \rightsquigarrow \mathcal{S}^{C^{op}}(A\times C,Y)$$

plays the role of Y^A. In particular if B^A exists in C for a pair of objects A,B in C then

$$(C)B^A \cong C(C,B^A) \cong C(A\times C,B) \cong \mathcal{S}^{C^{op}}(A\times C,B)$$

showing that the embedding preserves exponentiation.

<u>Theorem</u> Let A,Y be any objects in any category with finite products (including the empty product 1); then the following two statements cannot both be true

a) there exists $\bar{g}:A\times A \to Y$ such that for all $f:A \to Y$ there exists $x:1 \to A$ such that for all $a:1 \to A$

$$\langle a,x \rangle \bar{g} = a.f$$

b) there exists $t:Y \to Y$ such that for all $y:1 \to Y$

$$y.t \neq y.$$

Proof: Apply above remark and the proof in the previous section.

Of course the "transcendental" proof just given is somewhat ridiculous, since the incompatibility of a) and b) can be proved directly just as simply as it was proved in the previous section under the more restrictive hypothesis on C. However we wish to take the opportunity to make some further remarks about the above canonical embedding of an arbitrary (small) category into a cartesian closed category $\bar{C}$ (let the latter denote the smallest full cartesian closed subcategory of $\mathcal{S}^{C^{op}}$ which contains C). One of the standard ways of embedding a structure into a higher-order structure is to consider "definable" functionals, operators, etc.; however this is difficult to oversee from a simple-minded point of view since it usually requires enumerating all possible definitions. On the other hand in many situations (e.g. functorial semantics of algebraic theories or functorial semantics of elementary theories if the elementary theories are complete) one has come to expect that natural transformations are identical with definable ones or at least a reasonable substitute for definable ones. The latter alternative seems to be at least partly true in the present case. Thus for example we are led to the following definition. If A,B,C,D are objects in a category C with finite products, a <u>natural operator</u>

$$B^A \xrightarrow{\Phi} D^C$$

shall be simply a natural transformation between the exponential functors of the (functors corresponding to the) given objects in $\mathcal{S}^{C^{op}}$ (hence in $\bar{C}$). in particular if $C = 1$ we would call a natural operator a natural functional. Note that 1 will not be a generator for all of $\mathcal{S}^{C^{op}}$ unless $C = 1$; however it might conceivably be so for $\bar{C}$, and we have a partial result in that direction. In fact, in the case that 1 is a generator for C itself, we can describe in more familiar terms what a natural operator is.

Recall that "1 is a generator for C" simply means that a morphism $f:X \to Y$ in C is determined by its "values" $x.f:1 \to Y$ for $x:1 \to X$. In that case it is sensible to call the elements of the set $C(1,X)$ of points of X also the <u>elements of X</u>. Then a function

$$C(1,X) \to C(1,Y)$$

is induced by at most one C-morphism $X \to Y$, and in case it is, we say by abuse of language that the function <u>is</u> a morphism of C.

<u>Proposition</u> Suppose that C is a category with finite products in which 1 is a generator, and that A,B,C,D are objects of C. Then

1) a natural operator

$$B^A \xrightarrow{\Phi} D^C$$

is entirely determined by a single function

$$C(A,B) \xrightarrow{1\Phi} C(C,D)$$

and

2) such a function determines a natural operator iff for every object X of C and for every C-morphism $f: A\times X \to B$, the function

$$C(1,C\times X) \xrightarrow{(f)(X\Phi)} C(1,D)$$

is a C-morphism, where $(f)(X\Phi)$ is defined by

$$\langle c,x \rangle \big((f)(X\Phi)\big) = (c)\big((f_x)(1\Phi)\big)$$

for any $c: 1 \to C$, $x: 1 \to X$, f_x denoting the composition

$$\begin{array}{ccccc} A & \longrightarrow & A\times X & \xrightarrow{f} & B. \\ & \searrow_{\cong} & \nearrow_{A\times x} & & \\ & & A\times 1 & & \end{array}$$

Proof: We are abusing notations to the extent of identifying a morphism with its λ-transform via the bijections of the form

$$C(A\times X,B) \cong \bar{C}(A\times X,B) \cong \bar{C}(X,B^A).$$

Actually the given operator Φ is a family of functions

$$C(X,B^A) \xrightarrow{X\Phi} C(X,D^C)$$

one for each object of C; the "naturalness" condition which this family must satisfy, is, via the abuse, that for every morphism $x: X' \to X$ of C, the diagram

$$\begin{array}{ccc} C(A\times X,B) & \xrightarrow{X\Phi} & C(C\times X,D) \\ \downarrow x & & \downarrow x \\ C(A\times X',B) & \xrightarrow[X'\Phi]{} & C(C\times X',D) \end{array}$$

should commute. Now let $X' = 1$. Since 1 is a generator for C, the value of the function $X\Phi$ at a given $f: A\times X \to B$ is determined by the knowledge, for each element x of X and element c of C, the result reached in the lower right hand corner by going across then down in the commutative diagram

$$\begin{array}{ccccc}
C(A\times X,B) & \xrightarrow{X\Phi} & C(C\times X,D) & \xrightarrow{c} & C(X,D) \\
\downarrow{\scriptstyle x} & & \downarrow{\scriptstyle x} & & \downarrow{\scriptstyle x} \\
C(A,B) & \xrightarrow[1\Phi]{} & C(C,D) & \xrightarrow[c]{} & C(1,D).
\end{array}$$

But since the same results are obtained by going down then across, all the functions $X\Phi$ are determined by the one function 1Φ, proving the first assertion. The second assertion is then clear, since the definition of $(f)(X\Phi)$ given in the statement of the proposition is just such as to assure naturality of $X\Phi$ provided its values exist.

To make the situation perfectly clear, notice that morphisms whose codomain is an exponential object can be discussed even though the exponential object does not exist, just by considering instead morphisms whose domain is a product. There is however then the problem of determining the morphisms whose domain is an exponential, and considering them to be the natural operators is in many contexts the smoothest and most "natural" thing to do. Experts on recursive functions or C^{∞} functions between finite-dimensional manifolds may wish to consider the result of taking C to be these particular categories in the above considerations. They may also wish to consider whether the fixed-point theorem of section one has any applications in those cases.

3. In order to apply the theorem of the previous section to obtain Tarski's theorem concerning the impossibility of defining truth for a theory within the theory itself, we first note briefly how a theory gives rise to a category C with finite products. Consider two objects A,2 and let the C-morphisms be equivalence classes of (tuples of) formulas or terms of the theory, where two formulas (or terms) are considered equivalent iff their logical equivalence (or equality) is provable in the theory. Thus the morphisms $1 \to A$ are (classes of) constant terms, the morphisms $A\times A \to A$ are (classes of) terms with two free variables, while morphisms $A^n \to 2$ are (classes of) formulas with n free variables so that in particular morphisms $1 \to 2$ are (classes of) sentences of the theory. In particular there is a morphism true: $1 \to 2$ corresponding to the class of sentences provable in the theory and similarly a morphism false: $1 \to 2$ corresponding to the class of sentences whose negation is provable in

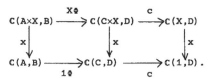

the theory. Morphisms $2^n \to 2$ would include all propositional operations, but we will make no use of that except for the following case:

If the theory is consistent there is a morphism not: $2 \to 2$ such that φ not $\neq \varphi$ for all morphisms $\varphi : 1 \to 2$

In particular we will not need to use the fact that $2 = 1+1$, although that determines the nature of those hom-sets not explicitly spelled out above. Defining composition to correspond to substitution (for example a constant $a : 1 \to A$ composed with a unary formula $\varphi : A \to 2$ composed with not gives the sentence $a\varphi$ not: $1 \to 2$, etc.) we get a category $\mathbb{C}$ with finite products which might be called the Lindenbaum category of the theory. Models of the theory can then be viewed as certain functors $\mathbb{C} \to \mathcal{S}$. We make no use here of the operation in $\mathbb{C}$ induced by quantification in the theory, but the categorical description of this operation will be clear to readers of the two papers cited in the introduction. In our construction above of $\mathbb{C}$ we have tacitly assumed that the theory was a first-order single-sorted one, in which case all objects of $\mathbb{C}$ are isomorphic to those of the form $A^n \times 2^m$, but with trivial modifications we could have started with a higher-order or several-sorted theory with no change of any significance to the arguments below. To make one point somewhat more explicit note that the projection morphisms $A^n \to A$ arise from the variables of the theory.

We then say that *satisfaction is definable* in the theory iff there is a binary formula $\text{sat} : A \times A \to 2$ in $\mathbb{C}$ such that for every unary formula $\varphi : A \to 2$ there is a constant $c : 1 \to A$ such that for every constant a the following diagram commutes in $\mathbb{C}$

$$\begin{array}{ccc} 1 & \xrightarrow{a} & A \\ {\scriptstyle \langle a,c \rangle}\downarrow & & \downarrow{\scriptstyle \varphi} \\ A \times A & \xrightarrow[\text{sat}]{} & 2 \end{array}$$

Here we imagine taking for c a Gödel number for (one of the representatives of) φ. The condition would traditionally be expressed by requiring that the sentence

$$a \text{ sat } c \longleftrightarrow a\varphi$$

be provable in the theory, but if $\mathbb{C}$ arises from our construction of the Lindenbaum category this amounts to the same thing.

Combining the above notion with our remark about the meaning of consistency and the theorem of the previous section we have immediately the

Corollary If satisfaction is definable in the theory then the theory is not consistent.

In order to show that Truth cannot be defined we first need to say what Truth would mean, which seems to require some further assumptions on the theory, which are however often realizable. Namely we suppose that there is a binary term

$$A\times A \xrightarrow{\text{subst}} A$$

in $\mathbb{C}$ and a ("metamathematical") binary relation

$$\Gamma \subseteq \mathbb{C}(1,A)\times\mathbb{C}(1,2)$$

between constants and sentences for which the following holds.

1) For all $\varphi:A \to 2$ there is $c:1 \to A$ such that for all $a:1 \to A$

$$(a \text{ subst } c)\,\Gamma(a\varphi)$$

For example we could imagine that $d\Gamma\sigma$ means that d is the Gödel number of some one of the sentences which represent σ, and that subst is a binary operation which, when applied to a constant a and to a constant c which happens to be the Gödel number of a unary formula φ, yields the Gödel number of the sentence obtained by substituting a into φ.

Given a binary relation $\Gamma \subseteq \mathbb{C}(1,A)\times\mathbb{C}(1,2)$ we say that Truth (of sentences) is definable in the theory (relative to Γ) provided there is a unary formula Truth:$A \to 2$ such that

2) For all $\sigma:1 \to 2$ and $d:1 \to A$, if $d\Gamma\sigma$ then $d\text{Truth} = \sigma$

Again the traditional formulation would require that the sentence

$$\ulcorner\sigma\urcorner \text{ Truth} \longleftrightarrow \sigma, \text{ for } \ulcorner\sigma\urcorner\Gamma\sigma$$

be provable, but in the Lindenbaum category this just amounts to the equation $\ulcorner\sigma\urcorner\text{Truth} = \sigma$.

Theorem If the theory is consistent and substitution is definable relative to a given binary relation Γ between constants and sentences, then Truth is not definable relative to the same binary relation

Proof: If both 1) and 2) hold then the diagram

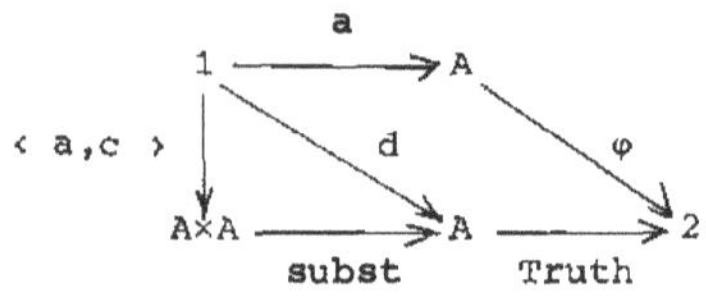

shows that

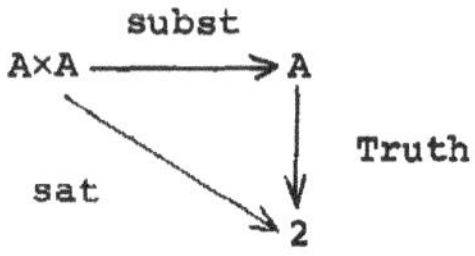

is a definition of satisfaction, contradicting the previous result.

We will also prove an "incompleteness theorem", using the notion of a Provability predicate. Given a binary relation Γ between constants and sentences, we say that <u>Provability is representable in the theory</u> iff there is a unary formula $Pr: A \to 2$ such that

3) Whenever $d\Gamma\sigma$ then

$$dPr = true \quad \text{iff} \quad \sigma = true$$

<u>Theorem</u> Suppose that for a given binary relation Γ between constants and sentences of C, substitution is definable and Provability is representable. Then the theory is not complete if it is consistent.

Proof: Suppose on the contrary that $C(1,2) = \{false, true\}$. Our notion of consistency implies that $false \neq true$. Condition 3) states that for $d\Gamma\sigma$

a) $\sigma = true$ implies $dPr = true$

b) $\sigma \neq true$ implies $dPr \neq true$

By completeness b) implies

b') $\sigma = false$ implies $dPr = false$

But a) and b') together with completeness mean that whenever $d\Gamma\sigma$

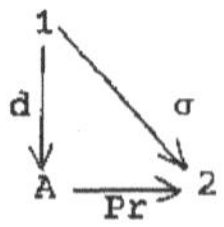

is commutative, i.e. that Pr satisfies condition 2) for a Truth-definition, which by our previous theorem yields a contradiction.

Note: Our proposition in section two can be interpreted as a fragment of a general theory developed by Eilenberg and Kelly from an idea of Spanier.

FOUNDATIONS FOR CATEGORIES AND SETS

by

Saunders Mac Lane*

I INTRODUCTION

A pressing problem confronting category theory is that of providing an adequate, precise, and flexible foundation. Two approaches are currently in use; neither is really satisfactory.

One current approach uses the class-set distinction provided by Gödel-Bernays axiomatic set theory. Here a small category is described as a set of morphisms equipped with an operation of composition satisfying the usual properties, while a large category is a class of morphisms with composition, and having the same properties. This approach does provide for the familiar large categories (the category of all sets, the category of all (small) groups, the category of all small categories) and for the functor category A^B for B small. However, it does not provide for the functor category A^B for B a large category.

Another approach uses Grothendieck's notion of a universe U — a set such that the elements $x \in U$ with the given membership relation $x \in y$ between such elements form a model of Zermelo-Fraenkel axiomatic set theory. A U-category is then a set x of morphisms with composition such that $x \in U$.

---

* University of Chicago. The investigations reported here were supported by an ONR grant.

This notion is essentially that of a small (better, a U-small) category; the category of all U-categories is then a U'-category for some larger universe U'. In this approach we may form functor categories, provided one adds to set theory the axiom that every set is a member of a universe. This is a strong axiom of infinity. Also, on this approach, no one to my knowledge has adequately examined the relation (say) between the category of all rings in one universe U and that of all rings in some larger universe.

Common to both approaches is the idea of using categories of different sizes (small and large, or in U, in U'). In effect, this is a use of categories which lie in different models of set theory (that is, in a "variable" set theory). Among axiomatic set theorists, it is a common dictum that there is only one set theory, though perhaps one not yet completely described by the usual axioms (Zermelo-Fraenkel plus axioms of extent). However, the standard developments of category theory concern repeatedly constructs such as the category of all groups, all sets, or all categories. The only visible approach to such constructs is some use of "variable" models of set theory.

Once the idea of a variable set theory is accepted, it becomes clear that the set theory need not be as strong as Zermelo-Fraenkel axiomatics require. For most mathematical purposes (ordinal numbers aside) Zermelo set theory is adequate. To define a category a very much weaker set theory suffices — a theory with elements and sets, but no sets of sets.

This paper is devoted to the investigation of one such weak axiomatic set theory. A model of these weak axioms will be called a "school" because it has classes. The chief development will be an abstract form of Freyd's adjoint functor theorem, in which the usual set-class distinction will be replaced by a more general subschool-school distinction.

II SCHOOLS

A school S is a collection of things, some of which are called <u>items</u> x, y, z, ..., while others are called <u>classes</u> A, B, C. Everything in the school S is either an item or a class, and possibly both. There is given a membership relation $x \in A$, to be read "the item x is an element of the class A", and there is an ordered pair operation which assigns to items x and y a new item $\langle x,y\rangle$ called the <u>ordered pair</u> of the given items x and y. These data are to be subjected to the following axioms:

<u>Extensionality for Classes</u>

If A and B are classes, and if for all items x, $x \in A$ if and only if $x \in B$, then $A = B$.

<u>Ordered Pair</u>

If $\langle x,y\rangle = \langle x',y'\rangle$, then $x = x'$ and $y = y'$.

Empty Class

There is a class $\emptyset$ with no items as members.

Unit Classes

To each item x there is class $\{x\}$ with x as its only element.

Cartesian Product

To classes A and B there exists a class $A \times B$ whose elements are all the ordered pairs $\langle x,y \rangle$ with $x \in A$ and $y \in B$.

Comprehension

If A is a class and $\psi(x)$ is a formula in which every bound variable is a variable item with a "limit" (see below), then there is a class C consisting of all those items x with $x \in A$ and $\psi(x)$.

The class C constructed in this axiom schema is the class usually written

$$C = \{x | x \in A \text{ and } \psi(x)\}.$$

Note that it is a subclass of the given class A and that the property ψ used to describe the elements of the subclass is to use only quantifiers which are "limited". By this we mean that each quantifier is to have one of the forms.

$(\exists y)\ y \in B$ --- or $(\forall z)\ z \in B$ implies ---

for some given class B, called the <u>limit</u> of the quantifier.

Ordered n-tuples $\langle x_1,\cdots,x_n\rangle$ of items may be defined by the usual recursion.

Using comprehension and cartesian product one can show for each natural number n that to classes $A_1,\cdots, A_n$ and a formula $\psi(x_1,\cdots,x_n)$ with all bound variables limited variables for items there exists a class

$$C = \{\langle x_1,\cdots,x_n\rangle \mid x_1 \in A_1,\cdots,x_n \in A_n \text{ and } \psi(x_1,\cdots,x_n)\}$$

Indeed, the cartesian product and comprehension axioms may be replaced by this stronger comprehension axiom for n-tuples.

Here are some examples of schools.

(1) Any realization ("model") of the Zermelo axioms for set theory with item = class = set (of the Zermelo model), and with the usual membership relation and the usual definition of ordered pair in terms of membership.

(2) If U is any set (in some set theory), form all the iterated cartesian products $U, U \times U$, $U \times (U \times U)$, $(U \times U) \times U,\cdots$ of U with itself, and take "class" to be any subset of one of these products and "item" to be any element of one of these products. The axioms for a school then hold. Note that this school is "homogeneous" in the "atoms" (elements u, $u' \in U$) in the following sense: There are classes of ordered pairs $\langle u,u'\rangle$ of atoms, and so on, but there need be no classes which simultaneously contain <u>both</u> ordered pairs and ordered triples. Such homogeneous set theories are indeed appropriate to

many mathematical purposes.

(3) If U, V, and W are given sets, (in some set theory) form all iterated cartesian products in which each of the factors is one of U, V, or W and take "class" to be any subset of one of these products and "item" to be any element of one of these products. This again is a homogeneous school.

(4) Let N be the set of natural numbers. Take any bijection $f: N \times N \longrightarrow N$ and define the ordered pair of $k, m \in N$ to be $\langle k,m \rangle = f(k,m)$. With these ordered pairs, take "item" to be any element of N and "class" to be any subclass of N; the axioms for a school hold.

Let S be a fixed school. A function can be defined as usual; thus if A and B are classes of S a function $F: A \longrightarrow B$ is a subset $F \subset A \times B$ with the familiar properties of a graph. By the comprehension axiom, we may construct the composite of two functions and prove it a function. The collection of all classes of the school, with all the functions as morphisms, then constitute a category, called the _category of the school_.

On the other hand, one may define what is meant by a _category within S_, following much the usual procedures. Such a category C consists of the following data

Two classes M (of morphisms) and O (of objects) in S.

Two functions $M \longrightarrow O$ giving domain and codomain of a morphism.

One function $M x_O M \longrightarrow M$ giving the composite of two morphisms.

Here $M x_0 M$ is the class of composable morphisms, described as usual in terms of domain and codomain; it exists in virtue of the comprehension axiom for a school.

These data are required to satisfy the usual axioms for a category.

This description thus indicates exactly how a category may be described adequately in a very weak set theory, such as the theory of a school. Functors and natural transformations between such categories within S may then be described in the evident ways.

Next we consider the axiom of choice for a school. The usual formulation of the existence of a choice function for a set of non-empty sets will not do, for there are no classes of classes in a school. Hence we say that S is a <u>choice school</u> when the following holds for any two classes A and B: If $H \subset A \times B$ is a class such that to each item $a \in A$ there exists an item $b \in B$ with $\langle a,b \rangle \in H$, then there is a function (i.e., a graph) $G \subset H$, such that for each $c \in A$ there is exactly one $b \in B$ with $\langle a,b \rangle \in G$. This form of the axiom of choice implies another known form: If $F: B \longrightarrow A$ has domain $B \neq \emptyset$, there is a function $G: A \longrightarrow B$ with $FGF = F$. To find this G, we simply apply the previous axiom to the set H of all those pairs $\langle a,b \rangle$ with $Fb = a$ or with $b \in B$ and a not in the image of F.

III NORMAL SUBSCHOOLS

A subschool T of a school S by definition consists of a collection, closed under the ordered pair operation, of some of the items of S and a collection of some of the classes of S such that these two subcollections themselves form a school (under the restriction of the membership relation of S). Thus if T is a subschool of S there is in T an empty class $\emptyset_T$, though this class $\emptyset_T$ need not be the original empty class $\emptyset$ of S; it is simply a class $\emptyset_T$ of S and of T none of whose elements in S are items of the subschool T. Similarly, the unit classes and the cartesian products of classes in the subschool T may differ from those in S.

We will restrict attention to subschools in which such pecularities do not arise. A subschool N of S is said to be normal when for every class A of N all those items x of S with $x \in A$ are items of N and all those classes B of S with $B \subset A$ in S are classes of N. These conditions imply that the empty class of S is the empty class of N, that the unit class $\{x\}$ in N of an item x of N is its unit class in S, and that the cartesian product $A \times B$ of two classes of N is their cartesian product in S. A similar remark holds for any class constructed by an application of the comprehension axiom.

A normal subschool N of S is completely determined

by the specification of which classes of S belongs to N, in the following sense.

Proposition 1

If S is a school, than a subcollection L of the collection of all classes of S is the collection of all the classes of a normal subschool N of S if and only if L contains the empty class of S, with any class of S all its subclasses in S and with any two classes A and B of S their cartesian product in S. When these conditions hold the subschool N is uniquely determined by the collection L. In detail, an item x of S is an item of N if and only if $x \in A$ for some class A of L or, equivalently, if and only if $\{x\}$ is in L.

The proof is left to the reader.

This notion of a normal subschool includes many basic examples, most notably the normal subschool composed of all the sets within a Gödel-Bernays universe of sets and classes. Indeed, let G be any Gödel-Bernays universe; that is a collection of classes, some of which are called sets, which satisfy the G-B axioms. Then G determines a school S_G in which the items are the sets of G, with the standard construction of ordered pairs, and the classes are the classes of G with the given membership relation. Since we have chosen to regard each set as a (special sort of) class, and since the empty class is a set, since the cartesian product of sets is a set and finally since every subclass of a set is a set, the collection L of all sets in any G-B school determines a normal subschool N — the subschool where both the items and the classes are the sets of the G-B universe G.

Let U be a Zermelo-Fraenkel universe; that is, a collection of sets which satisfy the usual Z-F axioms for set theory. Among the axioms is the axiom schema of replacement, which implies the usual axiom schema of comprehension. The weaker Zermelo axioms include the axiom schema for comprehension but not that for replacement. Consider a set $V \subset U$ such that V is a model for Zermelo set theory which is <u>transitive</u> ($x \in y \in V$ implies $x \in V$) and <u>inclusive</u> ($t \subset y \in V$ implies $t \in V$). Now U determines a school S_U with item = class = set of U, while the set V is the collection of classes for a normal subschool of S_U.

A Grothendieck universe will also provide a normal subschool. Suppose that the Z-F universe U satisfies strong axioms of infinity, so that there are in U sets W which are Grothendieck universes (i.e., transitive and inclusive sets W such that the elements $x \in W$ satisfy the Z-F axioms. It again follows that the set W is the collection of classes (and of items) for a normal subschool of the school S_U.

To motivate our study of completeness, let us observe the possible sizes of Zermelo universes V. We use the usual power set $P(x) = \{y|y \subset x\}$. Within U one may define the set R_α of all sets of rank less than the ordinal number α by the recursion

$$R_0 = \emptyset, \; R_{\alpha+1} = P(R_\alpha), \quad R_\beta = \bigcup_{\alpha<\beta} R_\alpha$$

the latter for β a limit ordinal, then the set $R_{\omega+\omega}$ is a Zermelo universe. Note that this universe does not contain the set $\{\omega, P\omega, P^2\omega, \ldots\}$ though the latter set must be present in any Z-F universe, because it can be obtained from the set ω by replacement (i.e., as the image of ω under the function $n \longmapsto P^n\omega$). For the same reason this Zermelo universe does not contain the union $\bigcup P^n\omega$ or the cartesian product $\Pi P^n\omega$. The latter observation (non-existence of a product) gives an indication of what completeness cannot mean in the category of all sets of a Zermelo universe.

IV SMALL COMPLETENESS

Consider a fixed normal subschool N of a school S, and call things (items, classes) small when they are things of N and large when they are things in S but not necessarily in N. A category C within S is small when both the classes M(C) and O(C) of morphisms and objects are small; it is locally small when for each pair c, c' of objects of C, the class $\hom_C(c,c')$ of all morphisms $c \longrightarrow c'$ is small. A small category is then necessarily locally small, but not conversely. In particular, when N is the subschool of all sets in the school S_G of some G-B universe G, as described above, this usage of "small" and "locally small" is precisely the familiar one.

If I is any "index" category (a category within S) and $F: I \longrightarrow C$ is any functor, the notion of a limit of F is defined by the usual universal properties. We will say that F is a <u>small functor</u> when the category I, the image of the object function of F, and the image of the mapping function of F are all three small. The category C is <u>small complete</u> if and only if every small functor $F: I \longrightarrow C$ has a limit in C. In particular, a <u>small family</u> of objects of C is a function $A: I \longrightarrow O(C)$ where both the set I and the image of the function A are small, and a product object ΠA_i, for $i \in I$, of such a family is defined as usual; there are no smallness conditions imposed on the function $P: I \longrightarrow M(C)$ which gives the projections $P_i: \Pi A_i \longrightarrow A_i$ of the product. Similarly, a family of coterminal morphisms of C is a function $F: I \longrightarrow M(C)$ together with two objects a and b of C such that for each $i \in I$, F_i is a morphism $F_i: a \longrightarrow b$; the equalizer of such a family is defined as usual.

This definition of small completeness is so formulated as to include the category of all small classes in a Zermelo universe. Indeed, let S_U be the school of a Z-F universe and N_V the normal subschool defined from a model V for the Zermelo axioms, as described above. Then the category Ens_V of all small classes has V as its class of objects and the class of all functions between small classes as its class of morphisms; it is a large category which is locally small. Moreover, Ens_V is small complete. For a small family $A: I \longrightarrow V$ of objects has <u>both</u> I and $Im(A)$ small; hence the union $\cup A_i$ of all classes A is small, because it is a union of a small class of small classes (and the Zermelo axioms provide for the existence of such unions). The product ΠA_i is then described as usual as the class of all those functions $f: I \longrightarrow \cup A_i$ with $f_i \in A_i$ for each $i \in I$. Since this is also a small class, Ens_V does have all small products. A simpler argument shows that it has all equalizers. Its small completeness then follows from the following proposition, which translates to our general situation some familiar facts about the construction of limits.

<u>Proposition 2</u>

Let C be a category within S in which every object is small. If C has a product for every small family of its objects and an equalizer for every coterminal pair of its morphisms, then C has an equalizer for every small family of coterminal morphisms.

<u>Proof</u>. If $F: I \longrightarrow M(C)$ is a small family of morphisms $F_i: a \longrightarrow b$, then b small by hypothesis implies $\{b\}$ small. The constant function $I \times I \longrightarrow \{b\}$ is

therefore a small family of objects of C. Therefore, there is a product Πb, $I \times I$ times, and two evident maps $a \longrightarrow \Pi b$ whose equalizer is the equalizer of the given family F.

<u>Proposition 3</u>

Let S be a choice school. If C is a category which has a product for every small family of its objects and an equalizer for every small family of coterminal morphisms, then C is small complete.

Given a category I and a small functor $F: I \longrightarrow C$, the usual proof forms first the product ΠF_i over all objects i of I. Each morphism $f: j \longrightarrow k$ of I then gives two maps $\Pi F_i \longrightarrow F_k$; one chooses an equalizer for each such pair; the limit of the given F is then obtained as the intersection of these equalizers. It is this use of choice which requires the assumption of our proposition that S be a choice school. This assumption could be replaced by a hypothesis on C giving to each pair of coterminal morphisms a fixed equalizer.

A functor $U: C \longrightarrow X$ between two categories within S is said to be <u>small continuous</u> when every product in C of a small family of objects of C is mapped by U to a product in X of the image family (which is not required to be small in X) and when every equalizer in C of a pair of coterminal morphisms of C is mapped by U to an equalizer of the image pair. The standard argument shows for C small complete that a small continuous functor U carries the limit in C of any small functor $F: I \longrightarrow C$ into a limit of UF.

V THE ADJOINT FUNCTOR THEOREM

Again, N is a normal subschool of a choice school S.

Theorem 4

If C is a small complete and locally small category within S, X a category within S, and $U: C \longrightarrow X$ a small-continuous functor, then U has a left adjoint if and only if it satisfies the solution set condition.

The solution set condition asserts that to each object x of X there is a small class I and functions

$$f: I \longrightarrow M(X),\ b: I \longrightarrow O(X),\ f_i : x \longrightarrow U(b_i)$$

for $i \in I$ with $Im(f)$ and $Im(b)$ small such that to every morphism $g: x \longrightarrow U(a)$ there exist $i \in I$ and $g': b_i \longrightarrow a$ with $g = (Ug')f$. This is just Freyd's solution set condition, with the added requirements that $Im(f)$ and $Im(b)$ are small.

The proof of this theorem is the standard one, which we repeat now chiefly to put down in print certain simplifications well known in the folklore (Lawvere, Kelly, Freyd and others). The solution set condition is clearly necessary, since a functor F left adjoint to U produces a solution set with just one object $b = F(x)$ and one (universal) map $x \longrightarrow UF(x)$.

Conversely, let the solution set condition be given. By completeness, the small product $d = \Pi b_i$ with projections $p_i: d \longrightarrow b_i$ exists in C. By continuity, $Up_i: Ud \longrightarrow Ub_i$ is a product in X; there is therefore a morphism $f: x \longrightarrow Ud$ defined by $(Up_i)f = f_i$ for each $i \in I$. Given any $g: x \longrightarrow Ua$ as in the solution set condition, then there exists $g^*: d \longrightarrow a$ with $g = (Ug^*)f$. Since C is locally small, the class of all endomorphisms of the object d in C is small, and so is the class

$$K = \{k \mid k: d \longrightarrow d \text{ in } C \text{ with } (Uk)f = f\} \quad .$$

By completeness, there is an equalizer $e: d_0 \longrightarrow d$ for this class K of coterminal morphisms. Like any equalizer, e is monic.

<u>Lemma</u> (<u>Kelly</u>)

If $h: d \longrightarrow d_0$ has $U(eh)f = f$, then $he = 1$.

<u>Proof</u>. Both 1_d and eh are in K; since e is the equalizer of K, $ehe = e$; since e is monic, $he = 1$.

Because U is continuous, Ue equalizes all Uk with $(Uk)f = f$. But f is another equalizer of these Uk; hence there exists $m: x \longrightarrow Ud_0$ with $f = (Ue)m$. To any $g: x \longrightarrow Ua$ we now have $g = U(g^*)f$ and therefore $g = U(v)m$ for some $v: d_0 \longrightarrow a$; to show m universal it remains only to show this v unique. Suppose not, so $w: d_0 \longrightarrow a$ also has $U(w)m = g$. Take the equalizer $e_1: d_1 \longrightarrow d_0$ of $v, w: d_0 \longrightarrow a$. By continuity, Ue_1 equalizes Uv, Uw, and so does m. Therefore, $m = (Ue_1)m_1$ for some $m_1: x \longrightarrow U(d_1)$. As for any such morphism $g = m_1$, one then has $m_1 = U(s)f$ for some $s: d \longrightarrow d_1$, $d \xrightarrow{s} d_1 \xrightarrow{e_1} d_0 \xrightarrow{e} d$. But then e_1s has $U(e_1s)f = m$, so $U(ee_1s)f = f$ and by the lemma $e_1se = 1$. The (monic) equalizer e_1 thus has a right inverse, so is an isomorphism. Since e_1 was the equalizer of v and w, we have $v = w$.

To each object x we thus have an object d_0 of C and a universal morphism $m: x \longrightarrow Ud_0$. Choose one such d_0 for each x; then as usual $Fx = d_0$ is the object function of a functor F left adjoint to U.

In place of the axiom of choice, one might assume that every family of coterminal morphisms has a chosen equalizer.

This "abstract" version of the adjoint functor theorem shows that this theorem has several different sorts of interpretations: the usual one, where "small" means a set of G-B set theory, another where "small" means within a Grothendieck universe, and still another where "small" means within a chosen model of Zermelo set theory. The leading idea is that completeness conditions can be spelled out as completeness for certain specified limits. It seems probable that other theorems of category theory may be usefully ramified when overall "completeness" hypothesis are replaced by completeness for certain specified limits.

VI TYPES AND ITERATED SCHOOLS

The major problem of the foundations of category theory remains that of handling successfully larger categories "of all so-and-so's". This can be done effectively by using a succession of schools, like a cumulative type theory. First, let U be some set-theoretical universe, regarded as a school S with item = class = set of U. Construct a larger school S^1 with item = set of U, with the ordered pairs those given in U, and with class any subcollection of U. Thus S^1 is related to S as GB-class to G-B sets. Categories within S are small; those within S^1 are large. Now form a still larger school S^2 with items all (suitably constructed) n-tuples of classes of S^1, with U^2 the collection of all these items, and with class (in S^2) any subcollection of U^2. Categories within S^2 might be called giant. Continuation of this process yields a sequence of schools $S \subset S^1 \subset S^2 \subset S^3 \subset \ldots$, each a normal subschool of the next. The categories within S^n for n = 2 are exactly the "metacategories" frequently used in informal arguments. Such a scale of schools (a type theory on top of a set theory) may be described by a suitable axiom system. The standard proof that G-B set theory is consistently relative to ZF set theory (J. B. Rosserand Hao Wang,

"Non-standard Models for Formal Logics," *J. Symbolic Logic*, 15; 113-129, (1950)) can probably be translated to show that the consistency of the original set theory implies the consistency of a suitable language for any one S^n. In this way, the use of categories within schools can provide a suitable and consistent foundation for category theory, by providing an explicit setting for successively larger types of metacategories.

ON THE DIMENSION OF OBJECTS AND CATEGORIES III

HOCHSCHILD DIMENSION*

by

Barry Mitchell**

R. Swan has observed that the situations encountered in [7] and [8] can be considered as special cases of the following. Let K be a commutative ring, and let Λ be a K-algebra. (All rings have identities and all ring homomorphisms preserve identities.) Suppose that A is an additive category and let $\phi: K \longrightarrow C(A)$ be a ring homomorphism where $C(A)$ is the ring of endomorphisms of the identity functor on A. Let ${}^{\Lambda}A$ denote the category of left Λ-objects in A (that is, ring homomorphisms $\Lambda \longrightarrow \mathrm{Hom}_A(A,A)$ with $A \in A$). Let ${}^{\Lambda}A_{\phi}$ denote the full subcategory of ${}^{\Lambda}A$ consisting of all those Λ-objects such that the composition $K \longrightarrow \Lambda \longrightarrow \mathrm{Hom}_A(A,A)$ is the same as the homomorphism induced by ϕ. If A is abelian, then so is ${}^{\Lambda}A_{\phi}$.

As an example, consider a ring homomorphism $\phi: K \longrightarrow C(A)$ and let $f(x)$ be a polynomial with coefficients

---

* Research supported by National Science Foundation Grant No. GP-6024.

** Dept. of Math., Bowdoin College, Brunswick, Maine.

in K. Then the category A_f of all endomorphisms α in A satisfying $f(\alpha) = 0$ (see [7, §4]) is the same as the category ${}^{\Lambda}A_{\phi}$ where Λ is the K-algebra $K[x]/(f(x))$. Another example is had by replacing A with ${}^{\Gamma}A$ where Γ is another K-algebra and taking $\phi: K \longrightarrow C({}^{\Gamma}A)$ to be the obvious ring homomorphism. Then ${}^{\Lambda}({}^{\Gamma}A)_{\phi}$ is the category of all objects in A which have simultaneous Λ and Γ operations such that Λ-operations commute with Γ-operations and both induce the same K-operations. This is the same as the category ${}^{\Lambda\otimes\Gamma}A$ of left $\Lambda\otimes\Gamma$ objects (tensor product over K).

Swan has further observed that lemma 1.4 of the present paper can be used to give an upper bound for the global dimension of ${}^{\Lambda}A_{\phi}$ in the case where the Hochschild dimension of Λ (dim Λ) is known. The latter is defined as the homological dimension of Λ considered as a $\Lambda^e = \Lambda^* \otimes \Lambda$ module. Actually we shall be using this lemma inversely to determine a lower bound for dim Λ in certain cases where gl. dim. ${}^{\Lambda}A_{\phi}$ is known.

Sometimes it is true that

$$\text{h.d. } \Lambda \otimes_K A = d + \text{h.d.}_A\, A \qquad (1)$$

for all non-zero objects A in an abelian category A, where $d = \dim \Lambda$ and $\Lambda \otimes_K A$ is considered as an object of the category ${}^{\Lambda^e}A_{\phi}$.

This question is of particular interest when Λ is of the form $K(\pi)$, π denoting a finite partially ordered set. The question is then related to the existence of a "dimension" for a finite ordered set or, in other words, an integer d' such that

$$\text{gl.dim. } {}^{\pi}A = d' + \text{gl.dim. } A$$

for all abelian categories A. Here ${}^{\pi}A$ denotes the category of covariant functors from π to A. For $d' = 0, 1,$ or 2, we shall show that $d = d'$ and that equation (1) is always valid. However it is now known that not every finite partially ordered set has a dimension d'. A counter-example involving an ordered set of 15 elements has been produced by William T. Spears at the University of Florida. Details will appear in his thesis.

Throughout this paper A will denote an abelian category.

1. AN ADJOINT RELATION

Let A be an (abelian) category with coproducts indexed by some cardinal number ρ. If Λ is a ring and M is a right Λ-module, then the tensor product $M \otimes_{\Lambda} A$ is defined for all left Λ-objects A in A, providing M admits an exact sequence of right Λ-modules

$$J_\Lambda \longrightarrow I_\Lambda$$

where I and J have cardinal numbers no bigger than ρ. When using the above tensor product we shall always assume that M satisfies this condition without always stating so. We then have a natural equivalence of trifunctors

$$\mathrm{Hom}_\Lambda(M, \mathrm{Hom}_{\mathcal{A}}(A,B)) \approx \mathrm{Hom}_{\mathcal{A}}(M \otimes_\Lambda A,B) \tag{1}$$

for $M \in \mathcal{G}^\Lambda$, $A \in {}^\Lambda\mathcal{A}$, and $B \in \mathcal{A}$ (see [9,§3]). Here $\mathcal{G}$ denotes the category of abelian groups.

If M is projective and the coproducts in $\mathcal{A}$ are exact, then $M \otimes_\Lambda A$ is an exact functor of the Λ-object A. This is trivial for $M = \Lambda$ and, consequently, for M free because of the exact coproducts. That it is true for all projective M then follows from the fact that any projective is a retract of a free.

Lemma 1.1

Let $\phi\colon K \longrightarrow C(\mathcal{A})$ be a ring homomorphism and let Λ and Γ be K-algebras. Then there is a natural equivalence of trifunctors

$$\mathrm{Hom}_{\Lambda\otimes\Gamma^*}(M, \mathrm{Hom}_{\mathcal{A}}(A,B)) \approx \mathrm{Hom}_\Gamma(M \otimes_\Lambda A,B) \tag{2}$$

for $M \in \mathcal{G}^{\Lambda\otimes\Gamma^*}$, $A \in {}^\Lambda\mathcal{A}_\phi$, and $B \in {}^\Gamma\mathcal{A}_\phi$.

Proof. Since Λ and Γ induce the same K-operators on $\mathrm{Hom}_{A}(A,B)$, it follows that the latter is a right $\Lambda \otimes \Gamma^*$-module, and so the left side of (2) makes sense. Now, forgetting Γ-operators, we have the natural equivalence (1), and using naturality in M and B, we see that the Γ-morphisms on either side correspond to each other. This yields (2).

Corollary 1.2

Relative to K-algebras Λ, Γ, and Σ, there is a natural equivalence of trifunctors

$$\mathrm{Hom}_{\Lambda\otimes\Gamma^*}(M, \mathrm{Hom}_{\Sigma}(A,B)) \approx \mathrm{Hom}_{\Sigma\otimes\Gamma}(M \otimes_{\Lambda} A,B)$$

for $M \in G^{\Lambda\otimes\Gamma^*}$, $A \in {}^{\Lambda\otimes\Sigma}A$, and $B \in {}^{\Sigma\otimes\Gamma}A$.

Proof. Replace A by ${}^{\Sigma}A$ in 1.1.

Corollary 1.3

Let $\Lambda \longrightarrow \Gamma$ be a K-algebra homomorphism. Then the induced functor $T\colon {}^{\Gamma}A_{\phi} \longrightarrow {}^{\Lambda}A_{\phi}$ has as left adjoint the functor $S\colon {}^{\Lambda}A_{\phi} \longrightarrow {}^{\Gamma}A_{\phi}$ given by $S(A) = \Gamma \otimes_{\Lambda} A$, and as right adjoint the functor $R\colon {}^{\Lambda}A_{\phi} \longrightarrow {}^{\Gamma}A_{\phi}$ given by $R(A) = \mathrm{Hom}_{\Lambda}(\Gamma,A)$ (symbolic Hom).

Proof. The dual of 1.1 yields (interchanging the roles of Λ and Γ so as to get A and B in the right places)

$$\mathrm{Hom}_{\Lambda\otimes\Gamma^*}(M,\ \mathrm{Hom}_{A}(A,B) \approx \mathrm{Hom}_{\Lambda}(A,\mathrm{Hom}_{\Gamma}(M,B))\ . \qquad (3)$$

Then, taking $M = \Gamma$ (with right Λ-operators determined by the given algebra homomorphism) and combining (2) and (3), we obtain

$$\mathrm{Hom}_{\Gamma}(\Gamma \otimes_{\Lambda} A,B) \approx \mathrm{Hom}_{\Lambda}(A,\ \mathrm{Hom}_{\Gamma}(\Gamma,B))\ .$$

Since $\mathrm{Hom}(\Gamma,B)$ as a left Λ-object is just $T(B)$, this shows that S is the required left adjoint. That R is the right adjoint follows by duality.

<u>Remark 1.</u> The underlying assumption needed to obtain the left adjoint S (right adjoint R) is that A have exact coproducts (products) indexed by the maximum of the cardinal numbers of I and J, where Γ admits an exact sequence of right Λ-modules

$$J_{\Lambda} \longrightarrow I_{\Lambda} \longrightarrow \Gamma \longrightarrow 0\ .$$

If the coproducts (products) in A are exact and Γ is Λ-projective, then the left (right) adjoint is exact.

<u>Remark 2.</u> If $\Lambda = K$, then ${}^{\Lambda}A_{\phi} = A$. Thus corollary 1.3 gives us left and right adjoints for the forgetful functor $T\colon {}^{\Gamma}A_{\phi} \longrightarrow A$.

<u>Lemma 1.4 (Swan)</u>

Let Λ be a K-projective algebra and let $d = \dim \Lambda$.

If the coproducts in $\mathcal{A}$ are exact, then for each $A \in {}^{\Lambda}\mathcal{A}_{\phi}$ we have

$$\text{h.d. } A \leq d + \text{h.d.}_{\mathcal{A}} A .$$

Consequently

$$\text{gl. dim. } {}^{\Lambda}\mathcal{A}_{\phi} \leq d + \text{gl. dim. } \mathcal{A} .$$

<u>Proof</u>. Let

$$0 \longrightarrow P_d \longrightarrow \text{---} \longrightarrow P_1 \longrightarrow P_o \longrightarrow \Lambda \longrightarrow 0 \qquad (4)$$

be a projective resolution for Λ as a Λ^e-module. Since Λ is a projective K-module, it follows that $\Lambda^e = \Lambda^* \otimes_K \Lambda$ is a projective right Λ-module and, consequently, (4) splits as a sequence of right Λ-modules. Hence, for $A \in {}^{\Lambda}\mathcal{A}_{\phi}$, we obtain an exact sequence

$$0 \longrightarrow P_d \otimes_{\Lambda} A \longrightarrow \text{---} P_1 \otimes_{\Lambda} A \longrightarrow P_o \otimes_{\Lambda} A \longrightarrow A \longrightarrow 0$$

in ${}^{\Lambda}\mathcal{A}_{\phi}$. It suffices then to prove that $\text{h.d. } P \otimes_{\Lambda} A \leq \text{h.d.}_{\mathcal{A}} A$ for projective Λ^e-modules P. In view of the exactness of coproducts in $\mathcal{A}$, we need only consider the case $P = \Lambda^e$. But, in this case, we have

$$\Lambda^e \otimes_{\Lambda} A = (\Lambda \otimes_K \Lambda) \otimes_{\Lambda} A = \Lambda \otimes_K A = S(A)$$

where, by remark 2 following 1.3, S is the left adjoint for the forgetful functor $T: {}^{\Lambda}\mathcal{A}_{\phi} \longrightarrow \mathcal{A}$. Again, because of exact coproducts and projectivity of Λ as a K-module, S is exact, and so our conclusion follows from [7, 1.2].

Lemma 1.5

Let Λ be a K-projective K-algebra, and suppose that the coproducts in A are exact. If M is a K-projective left Λ-module such that h.d. M = m, then

$$\text{h.d. } M \otimes_K A \leq m + \text{h.d.}_{A} A$$

for all $A \in A$, where $M \otimes_K A$ is regarded as an object of ${}^{\Lambda}A_{\phi}$.

Proof. Consider a projective resolution of left Λ-modules

$$0 \longrightarrow P_m \longrightarrow \cdots \longrightarrow P_1 \longrightarrow P_0 \longrightarrow M \longrightarrow 0 \;. \qquad (5)$$

Since Λ and M are K-projective, (5) splits as K-modules and so we obtain an exact sequence

$$0 \longrightarrow P_m \otimes_K A \longrightarrow \cdots \longrightarrow P_1 \otimes_K A \longrightarrow P_0 \otimes_K A \longrightarrow M \otimes_K A \longrightarrow 0 \;.$$

It suffices then to prove that h.d. $P \otimes_K A \leq \text{h.d.}_{A} A$ for projective Λ-modules P. Again we need only consider the case $P = \Lambda$. But $\Lambda \otimes_K A = S(A)$ where S is the left adjoint for $T\colon {}^{\Lambda}A_{\phi} \longrightarrow A$ and so once more our conclusion follows from [7, 1.2].

Corollary 1.6

Under the hypothesis of 1.5 we have

$$\text{h.d. } \Lambda \otimes_K A \leq d + \text{h.d.}_A A$$

for all $A \in A$, where $\Lambda \otimes_K A$ is considered as an object in ${}^{\Lambda^e}A_\phi$ and $d = \dim \Lambda$.

Proof. Since Λ is K-projective, so is Λ^e, and so the conclusion follows by replacing Λ by Λ^e and M by Λ in 1.5.

2. FUNCTOR CATEGORIES

If π is a small category and K is a ring, then we let $K(\pi)$ denote the free K-module on the set of morphisms of π. $K(\pi)$ can be converted into a ring (not necessarily with identity) if multiplication is defined by the rule

$$\begin{aligned}(kx)(\ell y) &= (k\ell)(xy) \text{ when range } y = \text{domain } x, \\ &= 0 \text{ otherwise.}\end{aligned}$$

Here $k\ell$ denotes the product of two elements of K, and xy denotes the composition of two morphisms in π. If π is a finite category (that is, π has only a finite number of objects although it may have an infinite number of morphisms), and if its objects are 1, 2, ..., n, then $K(\pi)$

has an identity, namely $1_1 + 1_2 + \ldots + 1_n$. If, furthermore, K is commutative then $K(\pi)$ is a K-algebra. Such rings $K(\pi)$ have been considered by Lawvere [6].

<u>Lemma 2.1</u>

Let π be a finite category and let $\phi\colon K \longrightarrow C(A)$ be any ring homomorphism where K is a commutative ring. Then we have an equivalence of categories

$$T\colon {}^{\pi}A \longrightarrow {}^{\Lambda}A_{\phi}$$

where $\Lambda = K(\pi)$.

Proof. Given a covariant functor $D\colon \pi \longrightarrow A$ (in other words, an object of ${}^{\pi}A$) we define

$$T(D) = \bigoplus_{i=1}^{n} D(i) ,$$

where $1,2,\ldots,n$ are the objects of π. If $x\colon j \longrightarrow k$ in π, then we define the action of x on $T(D)$ to be the endomorphism $\bar{x}$ of $T(D)$ given by

$$\bar{x}\, u_i = 0 \quad \text{for} \quad i \neq j ,$$
$$\bar{x}\, u_j = u_k\, D(x)$$

where u_i denotes the ith coproduct injection for $\bigoplus_{i=1}^{n} D(i)$. This action determines uniquely an object of ${}^{\Lambda}A_{\phi}$. We leave it to the reader to define the behavior of T with respect to morphisms.

Now define a functor $S: {}^{\Lambda}\mathcal{A}_{\phi} \longrightarrow {}^{\pi}\mathcal{A}$ by taking $S(A)(i) = 1_i A$ (that is, the image of the action of 1_i on A). If $x: j \longrightarrow k$, we have

$$x(1_j A) = (x 1_j) A = (1_k x) A = 1_k (xA) \subset 1_k A ,$$

and so x induces a morphism

$$S(A)(x): S(A)(j) \longrightarrow S(A)(k) .$$

In this way, $S(A)$ becomes an object of ${}^{\pi}\mathcal{A}$. Again we leave it to the reader to define the behavior of S with respect to morphisms.

It is now easy to see that ST is naturally equivalent to the identity functor on ${}^{\pi}\mathcal{A}$. In showing that TS is naturally equivalent to the identity functor on ${}^{\Lambda}\mathcal{A}_{\phi}$ one makes use of the relations $1_i 1_j = 0$ for $i \neq j$, $1_i^2 = 1_i$, and $\sum_{i=1}^{n} 1_i = 1$ in $K(\pi)$ to see that any $K(\pi)$-object A is the coproduct of its sub-objects $1_i A$ in $\mathcal{A}$.

If π_1 and π_2 are finite categories and K is a commutative ring, then we have

$$K(\pi_1) \otimes K(\pi_2) \approx K(\pi_1 \times \pi_2) ,$$

where $\pi_1 \times \pi_2$ is the product of categories. In particular, using the relation $K(\pi)^* = K(\pi^*)$, we see that the enveloping algebra of $K(\pi)$ is given by

$$K(\pi)^e = K(\pi^* \times \pi) .$$

Now consider an object $A \in \mathcal{A}$, where we are given a ring homomorphism $\phi: K \longrightarrow C(\mathcal{A})$. Then $K(\pi) \otimes_K A$, considered as an object in ${}^{K(\pi)^e}\mathcal{A}_\phi$, corresponds under the equivalence of categories established in 2.1 to an object of ${}^{\pi^* \times \pi}\mathcal{A}$ which we shall denote by $U_\pi(A)$. Using the functor S of 2.1, one sees that $U_\pi(A)$ can be defined explicitly as follows. On objects, $U_\pi(A)$ is given by

$$U_\pi(A)(i,j) = {}^{\mathrm{Hom}(i,j)}A \qquad (1)$$

for $(i,j) \in \pi^* \times \pi$. If $(x,y): (i,j) \longrightarrow (i',j')$ in $\pi^* \times \pi$ (that is, $x: i' \longrightarrow i$ and $y: j \longrightarrow j'$ in π) then we have

$$U_\pi(A)(x,y)\ u_z = u'_{yzx}$$

where u_z denotes the zth injection for the coproduct (1). If we regard $U_\pi(A)$ as an object of ${}^{\pi}({}^{\pi^*}\mathcal{A})$, then we see that its value at an object $j \in \pi$ is just $S_j(A)$, where S_j is the left adjoint of the evaluation functor T_j: ${}^{\pi^*}\mathcal{A} \longrightarrow \mathcal{A}$. In view of [5, 1.2] it follows that $U_\pi(A)$ can be considered as a functor with π as domain and whose values all have homological dimension equal to $\mathrm{h.d.}_{\mathcal{A}}A$. This remark will be used to get upper bounds for $\mathrm{h.d.}U_\pi(A)$.

We leave it to the reader to verify the following lemmas.

Lemma 2.2

If π and τ are small categories then, under the obvious isomorphism of categories

$$^{\tau^*\times\tau}(^{\pi^*\times\pi}A) \approx {}^{(\tau\times\pi)^* \times (\tau\times\pi)}A \quad ,$$

the object $U_\tau(U_\pi(A))$ corresponds to the object $U_{\tau\times\pi}(A)$.

Lemma 2.3

If $\tau \longrightarrow \pi$ is an equivalence of small categories then, under the induced equivalence $^{\pi^*\times\pi}A \longrightarrow {}^{\tau^*\times\tau}A$, the object $U_\pi(A)$ is taken into the object $U_\tau(A)$.

3. MONOIDS

If π is a monoid (that is, a category with one object), then the left adjoint S for the forgetful functor $T\colon {}^\pi A \longrightarrow A$ is given by $S(A) = A(\pi)$ where $A(\pi)$ denotes the coproduct of π copies of A. Denoting the action on $A(\pi)$ of an element (morphism) $x \in \pi$ by X, we have

$$Xu_y = u_{xy} \tag{1}$$

for all $y \in \pi$. In this case we have $u_\pi(A) = A(\pi)$ where the left operations of π on $A(\pi)$ are given by (1) and the right operations are given by

$$X^*u_y = u_{yx} \quad .$$

Of course we must assume that $\mathcal{A}$ has coproducts indexed by π. If these coproducts are exact then S is exact and, consequently, by [7, 1.2] we have

$$\text{h.d.}_{\pi}\ A(\pi) = \text{h.d.}_{\mathcal{A}}\ A\ . \qquad (2)$$

If τ is a submonoid of π, then $Z(\tau) \subset Z(\pi)$, and so, subject to the existence of appropriate coproducts in $\mathcal{A}$, corollary 1.3 gives us a left adjoint S for the restriction functor $T\colon {}^{\pi}\mathcal{A} \longrightarrow {}^{\tau}\mathcal{A}$. If π has a subset B such that each element $x \in \pi$ admits a unique representation of the form $x = bt$ with $b \in B$ and $t \in \tau$, then $Z(\pi)$ is free as a right $Z(\tau)$-module. Consequently, if $\mathcal{A}$ has exact coproducts indexed by B, then S exists and is exact. If $1 \in B$, then the relation $xt \in \tau$ with $t \in \tau$ implies that $x \in \tau$. In this case we call τ a <u>left partitioning</u> submonoid of π. A <u>right partitioning</u> submonoid of π is one whose dual is a left partitioning submonoid of π. If τ is right partitioning and $\mathcal{A}$ has exact products indexed by B, then T has an exact right adjoint. In this case we have by [7, 1.2]

$$\text{h.d.}_{\tau}\ A \leq \text{h.d.}_{\pi}\ A \qquad (3)$$

for any left π-object A.

Examples of partitioning submonoids are the following:

(1) If π is a group and τ is a subgroup, then τ is both left and right partitioning.

(2) If τ and θ are monoids and $\tau \oplus \theta$ is their free product, then taking B to be 1 together with all words ending with an element of θ, we see that τ is a left partitioning submonoid of $\tau \oplus \theta$. Similarly, τ is right partitioning.

(3) If I is a set and I' and J are subsets, then the partially free monoid of type $(J, J \cap I')$ is a left and right partitioning submonoid of the partially free monoid of type (I, I'). (See [7, §2] for the definition.) Indeed, this is just a special case of example (2).

(4) If τ and θ are monoids, then, taking B to be the subset of $\tau \times \theta$ consisting of all pairs of the form (1,s) with $s \in \theta$, we see that $\tau = \tau \times 1$ is a left and right partitioning submonoid of $\tau \times \theta$.

(5) If τ_1 and τ_2 are left partitioning submonoids of π_1 and π_2 respectively, then $\tau_1 \times \tau_2$ is a left partitioning submonoid of $\pi_1 \times \pi_2$. In particular, if τ is a left and right partitioning submonoid of π, then $\tau^* \times \tau$ is a left and right partitioning submonoid of $\pi^* \times \pi$.

Lemma 3.1

If τ is a left and right partitioning submonoid of π, and if A has coproducts and exact products indexed by π, then

$$\text{h.d.}_{\tau^* \times \tau} A(\tau) \leq \text{h.d.}_{\pi^* \times \pi} A(\pi)$$

for all $A \in A$.

Proof. Since τ is left and right partitioning, the relation $t_1 x t_2 \in \tau$ with $t_1, t_2 \in \tau$ implies that $x \in t$. It follows that $A(\tau)$ is a retract of $A(\pi)$ as left $\tau^* \times \tau$-objects. Therefore, the result follows from (3) if we replace π by $\pi^* \times \pi$, τ by $\tau^* \times \tau$, and A by $A(\pi)$.

Theorem 3.2

If A has exact coproducts indexed by π and π is a free monoid or group on one generator, then

$$\text{h.d.}_{\pi^* \times \pi} A(\pi) = 1 + \text{h.d.}_A A \qquad (4)$$

for all non-zero $A \in A$. If π is partially free and A has exact products as well as exact coproducts indexed by π, then (4) is valid.

Proof. Let π be partially free on I generators. Then we have an exact sequence in ${}^{\pi}A$

$$0 \longrightarrow I_{A(\pi)} \xrightarrow{\beta} A(\pi) \xrightarrow{\alpha} A \longrightarrow 0 \qquad (5)$$

relative to any left π-object A [7, 2.1]. Using (2) this yields, in the case where $\mathcal{A}$ has exact coproducts over π,

$$\text{h.d.}_{\pi} A \leq 1 + \text{h.d.}_{\mathcal{A}} A .$$

In particular, replacing $\mathcal{A}$ by ${}^{\pi}\mathcal{A}$, A by $A(\pi)$, and π by π^*, and using (2), we obtain

$$\text{h.d.}_{\pi^* \times \pi} A(\pi) \leq 1 + \text{h.d.}_{\mathcal{A}} A .$$

To prove the other inequality we consider first the case where π is generated by a single element. In this case sequence (5) is a sequence of $\pi^* \times \pi$ objects where A is assigned identity operators. (If π has more than one generator the morphism β is not a $\pi^* \times \pi$ morphism.) Now, by [7, §2] we have

$$\text{h.d.}_{\pi} A = 1 + \text{h.d.}_{\mathcal{A}} A$$

for all non-zero A with identity operators. A second application of this yields

$$\text{h.d.}_{\pi^* \times \pi} A = 2 + \text{h.d.}_{\mathcal{A}} A .$$

This, together with the exact sequence (5), yields

$$\text{h.d.}_{\pi^* \times \pi} A(\pi) \geq 1 + \text{h.d.}_{\mathcal{A}} A .$$

The general case then follows from 3.1 (taking J to consist of a single element in example (3)).

Corollary 3.3

If π is the direct product of n free monoids or groups on one generator, and if $\mathcal{A}$ has exact coproducts over π, then

$$\text{h.d.}_{\pi^* \times \pi} A(\pi) = n + \text{h.d.}_{\mathcal{A}} A$$

for all non-zero $A \in \mathcal{A}$. The same is true for direct products of partially free monoids if $\mathcal{A}$ is assumed further to have exact products over π.

Proof. This follows by an easy induction, using 2.2.

Remark 1. Corollary 3.3 is true even if n is infinite. This follows from 3.1, using the fact that an infinite product of partially free monoids has a k-fold product of partially free monoids as a direct factor for each positive integer k.

Remark 2. Corollary 3.3 implies the well-known result

$$\dim K[X_1,\ldots,X_n] = n .$$

4. ABELIAN GROUPS

If π is a finite group of order m, and if $A \in \mathcal{A}$ is such that m_A is an isomorphism, then

$$h.d._{\pi}A = h.d._{A}A \tag{1}$$

for any π-object structure on A [7, 3.4]. Applying this again we then see that

$$h.d._{\pi^{*}\times\pi}A(\pi) = h.d._{A}A \ . \tag{2}$$

Now, suppose that G is a finitely generated abelian group of rank r. Then we can write $G = F \times \pi$ where F is a free abelian group of rank r, and π is a finite abelian group. Combining (2), 3.3, and 2.2, we then obtain:

Proposition 4.1

Let G be a finitely generated abelian group of rank r, and suppose that the order of the torsion subgroup of G is m. If $\mathcal{A}$ has exact countable coproducts, then

$$h.d._{G\times G}A(G) = r + h.d._{A}A$$

for all non-zero $A \in \mathcal{A}$ such that m_A is an isomorphism.

Before looking at bigger abelian groups we state without proof a generalization due to B. Osofsky [10] of a theorem of I. Berstein [3]. (See also Balcerzyk [2].)

Theorem 4.2

Let D be a directed set of cardinal number $\aleph_n$ for some integer $n \geq 0$, and let $\{R_i, \pi_i^j\}$ and $\{M_i, \mu_i^j\}$

be direct systems of rings and abelian groups such that M_i is a left R_i-module for each $i \in D$, and such that

$$\mu_i^j(rm) = \pi_i^j(r)\ \mu_i^j(m)$$

for $r \in R_i$ and $m \in M_i$. If R and M are the direct limits, then

$$\text{h.d.}_R M \leq n + 1 + \sup_i \text{h.d.}_{R_i} M_i$$

As an application of this theorem let G be an abelian group of order $\aleph_n$ and rank r. Then G is the directed union of its finitely generated subgroups G_i where i runs through a directed set of cardinality $\aleph_n$. If K is a ring then $K(G)$ is the directed union of its subrings $K(G_i)$. Now let P be the set of primes p such that G has p-torsion and suppose that p is invertible in K for each $p \in P$. Then each G_i is a finitely generated abelian group of rank at most r, and the order of its torsion subgroup is invertible in K. Hence, we have

$$\text{h.d.}_{K(G_i)} K \leq r ,$$

where K is regarded as a left $K(G_i)$-module with trivial operators. Consequently, by theorem 4.2, we obtain

$$\text{h.d.}_{K(G)} K \leq n + r + 1 ,$$

and so there is a projective $K(G)$ resolution of the form

$$0 \longrightarrow P_{n+r+1} \longrightarrow \cdots \longrightarrow P_1 \longrightarrow P_0 \longrightarrow K \longrightarrow 0 \qquad (3)$$

Theorem 4.3

Let G be an abelian group of order $\aleph_n$ and rank r and let $\mathcal{P}$ be the set of primes p such that G has p-torsion. If $\mathcal{A}$ has exact coproducts indexed by $\aleph_n$ then

$$\text{h.d.}_G A \leq n + r + 1 + \text{h.d.}_{\mathcal{A}} A \qquad (4)$$

for all G-objects A such that p_A is an isomorphism for all $p \in \mathcal{P}$. Consequently

$$\text{h.d.}_{G\times G} A(G) \leq n + r + 1 + \text{h.d.}_{\mathcal{A}} A \qquad (5)$$

for all such A in $\mathcal{A}$. If further $\mathcal{A}$ has exact products indexed by $\aleph_n$, then

$$\text{h.d.}_{G\times G} A(G) \geq r + \text{h.d.}_{\mathcal{A}} A \qquad (6)$$

for all $A \in \mathcal{A}$.

Proof. The assumption on A gives it a unique K-object structure where K is the ring of rational numbers whose denominators are products of primes in $\mathcal{P}$. Now if P is a projective left $K(G)$-module, and if $P\otimes_K A$ is regarded as a G-object in $\mathcal{A}$ with x acting as $x \otimes x$, then we know by [5, 3.1] that

$$\text{h.d.}_G P\otimes_K A \leq \text{h.d.}_{\mathcal{A}} A$$

Consequently, (4) follows from tensoring (3) over K with A. Replacing $\mathcal{A}$ by ${}^{G}\mathcal{A}$ and A by A(G) we obtain (5). Inequality (6) is a result of 3.1 and 3.3 using the fact that G has a free abelian subgroup of rank r.

Remark 1. Let K be a commutative ring of finite global dimension and let p be a prime which is not invertible in K. Taking π to be a cyclic group of order p, we know by [7, 3.4] that $\text{h.d.}_{\pi} A = \infty$ where A is any non-zero K-module with trivial π-operators such that $p_A = 0$. From 1.4 it therefore follows that

$$\dim K(\pi) = \infty . \tag{7}$$

Then, using 3.1, we see that (7) holds for any group π which has an element of order p.

Remark 2. Inequality (4) was proved by Balcerzyk [2] for categories of modules.

5. FINITE ORDERED SETS

Throughout this section π will denote a finite ordered set or, in other words, a finite category such that Hom (i,j) is either empty or consists of one element for each ordered pair (i,j) of elements of π. In the latter case we write $i \leq j$. Any such category is equivalent to one with the property that if $i \leq j$ and $j \leq i$, then $i = j$.

In view of lemma 2.3 we shall always assume that π has this property. Our terminology will then follow that of [8].

Recall that if $D \in {}^{\pi}A$ and $D(i) = 0$ for all i not in some final subset π' of π, then the homological dimension of D is the same as that of its restriction to π'. Furthermore, if A is the category of abelian groups (or any category with enough projectives), then D has a projective resolution every term of which is zero at i for $i \notin \pi'$. For this reason we shall sometimes not bother to distinguish between D and its restriction to π'. In particular, if the restriction of D to π' is the constant functor at A for some $A \in A$, then we shall refer to D as a constant functor (or diagram).

Let $\tau(\pi)$ denote the ordered subset of $\pi^* \times \pi$ consisting of all those pairs (i,j) such that $i \leq j$ in π. Then for $A \in A$ the object $U_\pi(A)$ is easily seen to be the constant diagram at A over $\tau(\pi)$ extended by 0's to the rest of $\pi^* \times \pi$. Since $\tau(\pi)$ is a final subset of $\pi^* \times \pi$ we shall sometimes denote the restriction of $U_\pi(A)$ to $\tau(\pi)$ also by $U_\pi(A)$.

An element (i',j') is a cover for an element (i,j) in $\tau(\pi)$ if and only if either $i = i'$ and j' is a cover for j, or $j = j'$ and i' is a co-cover for i.

An element (i,j) is maximal in $\tau(\pi)$ if and only if i is minimal and j is maximal in π. In particular $\tau(\pi)$ is terminal if and only if π is both initial and terminal. The minimal elements of $\tau(\pi)$ are the elements of the form (i,i) with $i \in \pi$.

Some examples of $\tau(\pi)$ are given in the following table.

π $\tau(\pi)$

(1)

(2)

(3)

(4)

(5)

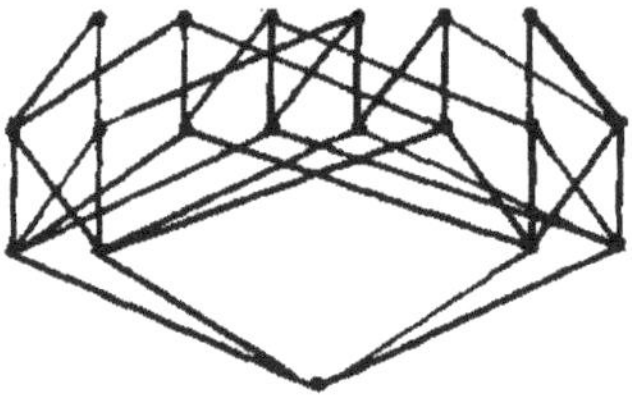
(6)

Before turning to the problem of computing the homological dimension of $U_\pi(A)$, we recall and apply two results from [8]. The following is theorem 3.4 of [8].

<u>Lemma 5.1</u>

Let D be pointwise free in ${}^{\pi}G$.

(a) If $\text{h.d.}_\pi\ D = m$ then $\text{h.d.}_\pi\ A \otimes D \leq m + \text{h.d.}_A\ A$ for all $A \in \mathcal{A}$.

(b) If there is an exact sequence in ${}^{\pi}G$

$$0 \longrightarrow K^{m-1} \longrightarrow P^{m-2} \longrightarrow \dots \longrightarrow P^{1} \longrightarrow P^{0} \longrightarrow D \longrightarrow 0 \quad (7)$$

with P^i projective and if K^{m-1} is split but not projective, then $\text{h.d.}_{\pi} A \otimes D \geq m + \text{h.d.}_{A} A$ for all nontrivial $A \in A$.

<u>Remark</u>. Part (a) is actually a special case of 1.5. The object $A \otimes D$, which is defined as the composition of $D: \pi \longrightarrow G$ with $A \otimes : G \longrightarrow A$, corresponds under the equivalence of categories T established in 2.1 to the $Z(\pi)$-object $A \otimes T(D)$.

In view of lemma 5.1 we say that a pointwise free object D in ${}^{\pi}G$ is <u>strong</u> if it has homological dimension m, and if it admits an exact sequence (7) with P^i projective and K^{m-1} split. The independence of this notion from the choice of the resolution (7) was established in [8] for special objects called muscles, but the proof (which is a simple application of Schanuel's lemma) applies to the general case. For such objects D we then have

$$\text{h.d.}_{\pi} A \otimes D = m + \text{h.d.}_{A} A$$

for all $A \in A$.

The following is lemma 3.7 of [8].

<u>Lemma 5.2</u>

If K^1 is the constant diagram at A in ${}^{\pi}G$ then there is an exact sequence

$$0 \longrightarrow K^2 \longrightarrow P^1 \longrightarrow K^1 \longrightarrow 0$$

where P^1 is projective and K^2 is split. In fact, we may take P^1 to be the diagram $\bigoplus_{m \in M} S_m(Z)$ where S_i denotes the left adjoint for the evaluation functor T_i and M denotes the set of minimal elements of π.

For $i \in \pi$ and $A \in \mathcal{A}$ we let $L_i(A)$ denote the object of ${}^{\pi}\mathcal{A}$ with A at i and 0 elsewhere. We then have an epimorphism $S_i(A) \longrightarrow L_i(A)$ in ${}^{\pi}\mathcal{A}$ whose kernel is the constant diagram at A over the set of elements of π strictly greater than i. Combining this with 5.2 we obtain:

<u>Corollary 5.3</u>

For any $i \in \pi$ we have an exact sequence in ${}^{\pi}\mathcal{G}$

$$0 \longrightarrow K^2 \longrightarrow P^1 \longrightarrow P^0 \longrightarrow L_i(Z) \longrightarrow 0$$

where P^1 and P^0 are projective and K^2 is split.

For a finite ordered set π we define

$$d(\pi) = \max_{i \in \pi} \text{h.d. } L_i(Z) .$$

<u>Lemma 5.4</u>

For any $D \in {}^{\pi}\mathcal{A}$ we have

$$\text{h.d.}_{\pi} D \leq d(\pi) + m ,$$

where $m = \sup_{i \in \pi} h.d.\ D_i$.

Proof. By induction on the number of elements of π . If π has only one element, then this is trivial. Otherwise let k be a minimal element of π , and form the exact sequence in ${}^{\pi}A$

$$0 \longrightarrow K \longrightarrow D \longrightarrow L_k(D_k) \longrightarrow 0 \qquad . \qquad (8)$$

Then $K_k = 0$, and $K_i = D_i$ for $i \neq k$. Now by 5.1 (a) we have

$$h.d.L_k(D) = h.d.D_k \otimes L_k(Z) \leq h.d._A D_k + h.d.L_k(Z) \leq d(\pi) + m .$$

Consequently if $h.d._{\pi} D > d(\pi) + m$, then from the exact sequence (8) we see $h.d._{\pi} K > m + d(\pi)$. But K can be considered as a diagram over $\pi - \{k\}$, and so this contradicts the induction.

Corollary 5.5

If $d(\pi) = 0, 1, 2,$ or 3, then

$$gl.\ dim.\ {}^{\pi}A = d(\pi) + gl.\ dim.\ A$$

for all nontrivial A. If $d(\pi) > 3$, then we have

$$3 + gl.\ dim.{}^{\pi}A \leq gl.\ dim.{}^{\pi}A \leq d(\pi) + gl.\ dim.A \quad (9)$$

Proof. The right hand inequality in (9) follows directly from 5.4 The other inequalities are consequences of 5.3 applied to 5.1(b).

Corollary 5.6

For any $A \in \mathcal{A}$ we have

$$U_\pi(A) \leq d(\pi) + \text{h.d.}_{\mathcal{A}} A \quad .$$

Proof. This follows from 5.4 in view of the fact that $U_\pi(A)$, regarded as an object of ${}^\pi({}^{\pi^*}\!\mathcal{A})$, has $S_i(A)$ at the ith vertex.

Corollary 5.7

If π has an element i such that h.d. $L_i(Z) = d(\pi)$ and $L_i(Z)$ is strong, then

$$\text{gl.dim.}\, {}^\pi\!\mathcal{A} = d(\pi) + \text{gl.dim.}\mathcal{A}$$

for all $\mathcal{A}$. In this case if K is a commutative ring of finite global dimension, then

$$\dim K(\pi) = d(\pi) \quad .$$

Proof. The first statement follows from 5.4 and 5.1 (b). Then taking $\mathcal{A}$ to be the category of K-modules, the other statement follows from 5.6 and 1.4.

Remark 1. Any generalized m-braid satisfies the hypothesis of 5.7 (See [8, §3]).

Remark 2. For any set π we have

$$d(\pi) - 1 \leq \dim Z(\pi) \leq d(\pi) \quad .$$

This follows from 5.6 and 1.4 in view of the fact that

$$\text{gl.dim.}\mathcal{G} = 1 \quad .$$

Theorem 5.8

If $d(\pi) = 0$, 1, or 2, then

$$h.d.U_{\pi}(A) = d(\pi) + h.d._{A}A$$

for all nontrivial $A \in \mathcal{A}$.

Proof. Corollary 5.6 gives us one inequality in any case. Now by corollary 5.5 and lemma 1.4 we have

$$d(\pi) \leq \dim Z(\pi) = h.d.U_{\pi}(Z)$$

for $d(\pi) = 0$, 1, 2, or 3. Therefore for such π we have

$$h.d.U_{\pi}(Z) = d(\pi) .$$

From 5.2 we then see that $U_{\pi}(Z)$ is strong when $d(\pi) = 0$, 1, or 2. Therefore our conclusion follows from 5.1.

By [8, 4.6] we know that

$$gl.\ dim.\ \mathcal{A} \geq 3 + gl.\ dim.\ A \qquad (10)$$

if and only if π contains a suspended crown as an uncrossed ordered subset. (See [8, §3] for the definitions.) Now it is easy to see that if π contains an uncrossed suspended crown, then it contains one of the form

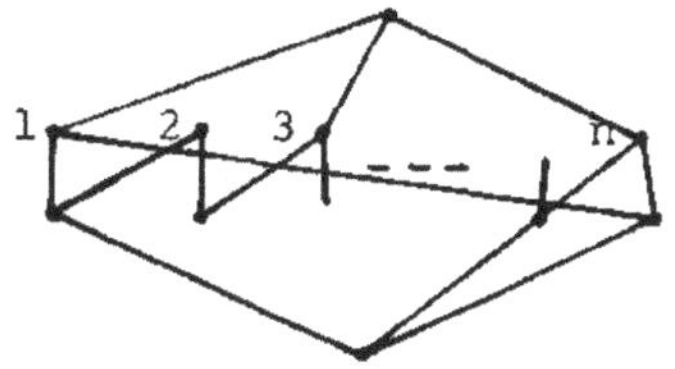

where $n \geq 2$. Let us denote this set by c_n. Thus c_2 is the same as (5), and $\tau(c_2)$ is (6). The set $\tau(c_n)$ is similar to $\tau(c_2)$, only wider. One can write down a projective resolution of the form

$$0 \longrightarrow P^3 \longrightarrow P^2 \longrightarrow P^1 \longrightarrow P^0 \longrightarrow U_{C_n}(Z) \longrightarrow 0$$

where the kernel of $P^1 \longrightarrow P^0$ is split but not projective. The matrices involved are rather complicated, and we shall not give any details here. It then follows from 5.1 that

$$h.d.U_{C_n}(A) = 3 + h.d._A A$$

for all nontrivial A.

REFERENCES

[1] Balcerzyk, S., "The Global Dimension of The Group Rings of Abelian Groups II", *Fund. Math.* 58; 67-73, (1966).

[2] Balcerzyk, S., "On Projective Dimension of Direct Limit of Modules". *Bull. Acad. Polon. Sci. Ser. Sci. Math. Astron. Phys.* 14; 241-244 (1966).

[3] Berstein, I., "On the Dimension of Modules and Algebras IX", *Nagoya Math. J.* 13; 83-84 (1958).

[4] Cartan, H. and S. Eilenberg, *Homological Algebra.* Princeton University Press, Princeton, New Jersey, 1956.

[5] Eilenberg, S., A. Rosenberg, and D. Zelinsky, "On the Dimension of Modules and Algebras, VIII". *Nagoya Math. J.* 12; 71-93, (1957).

[6] Lawvere, F. W., "The Group Ring of A Small Category". *Notices A.M.S.* 10, (April, 1963).

[7] Mitchell, B., "On the Dimension of Objects and Categories, I". *J. of Algebra* 9; 314-340, (1968).

[8] Mitchell, B., "On the Dimension of Objects and Categories, II". *J. of Algebra* 9; 341-368, (1968).

[9] Mitchell, B., "On Characterizing Functors and Categories". Mimeographed, Bowdoin College, (1967).

[10] Osofsky, B., "Upper Bounds on Homological Dimensions". (to appear.)

LOCALLY NOETHERIAN CATEGORIES AND GENERALIZED STRICTLY LINEARLY COMPACT RINGS. APPLICATIONS.

by

JAN-ERIK ROOS

(Department of Mathematics,
University of Lund, SWEDEN.)

CONTENTS

§ 0. Introduction.

The principal aim of this paper is to prove (and to give applications and precisions of) a structure theorem (cf. notably Theorem 6 and its corollaries below) for those categories $\underline{C}$ that satisfy the following two axioms:

A) $\underline{C}$ is an abelian category, that has direct limits that are exact functors when taken over directed sets (this is the axiom AB 5 of [30]);

B) $\underline{C}$ has a set of generators $\{N_\alpha\}_{\alpha \in I}$, where each N_α is a noetherian object of $\underline{C}$ (i.e. N_α satisfies the ascending chain condition on subobjects).

Following Gabriel [24], [25], we will say that $\underline{C}$ is a locally noetherian (abelian) category if A) and B) are verified.

Here are some examples of locally noetherian categories:

1) Mod(A) = the category of unitary left modules over a left noetherian ring A;

2) The category Qcoh(X) of quasi-coherent sheaves over a noetherian prescheme X [25];

3) The category $\text{Mod}(\underline{O}_X)$ of sheaves of $\underline{O}_X$-modules, where X is a locally noetherian prescheme with structure sheaf $\underline{O}_X$ [32];

4) The category $\text{Tors}(\mathbb{Z})$ of abelian torsion groups, and more generally any closed subcategory [25] of Mod(A), where A is as in 1);

5) The category of topological discrete G-modules, where G is a profinite group [58];

6) Let $A = (A^n)_{n \geq 0}$ be a graded ring, where A^o is a left noetherian ring, and where each A^n is a finitely generated left A^o-module for its natural left A^o-module structure. Then the category $\text{Modgr}^-(A)$ of negatively graded left A-modules is locally noetherian. However, the category $\text{Modgr}^+(A)$ of positively graded A-modules is locally noetherian if and only if A is noetherian in the graded sense, and this condition is strictly stronger than the one mentioned above (example: the Steenrod algebra, which is coherent, but not noetherian [15]);

7) The dual category of the category of pro-algebraic commutative group schemes over an algebraically closed field [44].

It should be remarked that the cases 2) and 3) are quite different and that in the examples 2)-7) the category $\underline{C}$ is not in general equivalent to a module category (i.e. a category of type 1)).

We make no attempt to summarize the results of this paper here, and we only remark that the principal result is that the study of locally noetherian categories is entirely equivalent to the study of a certain class of topological rings with a linear topology that is explicitly described. Thus, problems and results that seem to belong to "pure category theory" are in fact equivalent to "purely ring-theoretical" ones, and this gives new results in both directions. More details about this are in particular given in § 5, where it is also explained how our results are inspired by and related to earlier results of Kaplansky [35], Matlis [40], Grothendieck [31], Serre [56], Gabriel [24], [25] and Leptin [38], [39]. Several ring-theoretical and "categorical" applications as well as several open problems are scattered throughout the paper. Among the "categorical" applications, we wish particularly to stress those given in § 8, where we are able to classify all "exact sequences of categories"

$$0 \longrightarrow \underline{D} \longrightarrow \underline{C} \longrightarrow \underline{C}/\underline{D} \longrightarrow 0$$

where $\underline{D}$ is a localizing subcategory of $\underline{C}$ (locally noetherian), and where $\underline{D}$ is supposed to be stable under the formation of injective envelopes in $\underline{C}$.

Some results of this paper have been summarized in [54], and the present paper will be a part of a systematic study of Grothendieck categories, that we hope to publish soon (hopefully together with a solution of at least some of the problems mentioned above).

In order to facilitate the reading of this paper and in order to give some general background, we will first briefly review and complete some well-known results about locally noetherian categories.

§ 1. Characterizations of locally noetherian categories.

A category satisfying the axiom A) above, as well as the following weakened form of B:

B_g) $\underline{C}$ has a set of generators $\{G_\alpha\}_{\alpha \in I}$,
is generally called a Grothendieck category [27], and Grothendieck proved in [30] that such a category has sufficiently many injective objects. If now $\underline{C}$ is also locally noetherian, then we have much more precise results. In fact, Matlis proved in [40], that if $\underline{C}$ is of the form Mod(A), where A is a left noetherian ring, then every injective in $\underline{C}$ is isomorphic to a sum of indecomposable injective objects (which in the commutative case correspond to the prime ideals of A), and any two such decompositions are isomorphic in a natural way. Furthermore, in this case every sum and more generally every directed $\varinjlim$ of injectives in Mod(A) is again injective [12]. These results were extended by Gabriel [23], [24], [25], to the general locally noetherian case. However, there are also several converses of these results, which in the module case are due to Bass [4], Papp [45] and Faith-Walker [21] and which together with the Matlis results can be formulated as follows:

PROPOSITION 1.- *The following conditions on a category* Mod(A) *are equivalent*:

(i) Mod(A) *is locally noetherian* (i.e. A is left noetherian);

(ii) *Every injective object of* Mod(A) *is isomorphic to a direct sum of indecomposable injectives*;

(iii) *Any sum of injectives in* Mod(A) *is injective*;

(iv) *Any* (directed) *direct limit of injectives in* Mod(A) *is injective*;

(v) [Bounded decompositions of injectives.] *There exists a cardinal* c *such that every injective in* Mod(A) *is isomorphic to a sum of injectives, each having* $\leq c$ *generators*;

(vi) [Strict cogenerator.] *There exists an object* C *of* Mod(A) *such that every object in* Mod(A) *is isomorphic to a subobject of a direct sum of copies of* C.

If we now replace Mod(A) by a general Grothendieck category $\underline{C}$, then the conditions (i)-(vi) above are not all equivalent in general (as for the general formulation of (v), cf. below), a counter-example to the implications (iii) => (i), (iv) => (i), (vi) => (i) is for example given by a non-discrete spectral category [1]). The results of Gabriel are exactly the implications (i) => (ii), (i) => (iii) and (i) => (iv). Also the implication (ii) => (vi) is universally valid but we do not know for example whether (ii) => (i) is always true for all Grothendieck categories. As for the condition (v) as it stands it has no meaning for Grothendieck categories. However, if $\underline{c}$ is an infinite cardinal, it is easy to see that an A-module M has $\underline{<c}$ generators if and only if the functor $\mathrm{Hom}_A(M,\cdot)$ commutes with $\underline{c}$-directed unions. [For more details, and for relations with the notion of object of (abstract) finite type, we refer the reader to [53].] Thus, if we say that an object C in a Grothendieck category $\underline{C}$ has $\underline{<c}$ generators [$\underline{c}$ infinite cardinal, or the finite cardinal, cf. [53]] if the functor $\mathrm{Hom}_{\underline{C}}(C,\cdot)$ commutes with $\underline{c}$-directed unions, then (v) makes sense, and we can in fact prove the following

THEOREM 1.- The following conditions on a Grothendieck category $\underline{C}$ are equivalent:

(i) ["Bounded decompositions of injectives."] There exists a fixed cardinal $\underline{c}$ such that every injective I in $\underline{C}$ admits a direct decomposition $I = \coprod I_i$, where each I_i has $\underline{<c}$ generators;

(i)' There exists an injective object I such that every injective in $\underline{C}$ is a direct sum of direct factors of I;

(ii) Every sum of injectives in $\underline{C}$ is still injective;

(iii) ["Strict cogenerator."] There is an object D of $\underline{C}$, such that every object of $\underline{C}$ is a subobject of a suitable direct sum of copies of D.

---

[1]) A spectral category is a Grothendieck category, where every object is injective; such a category is locally noetherian if and only if it is discrete in the sense of [27]. For the general theory of spectral categories, and examples of non-discrete ones, we refer the reader to [51] and the literature cited there.

This result will not be needed in what follows, so we give no proof. Indications of the proof together with the necessary generalizations to Grothendieck categories of results of the Kaplansky type about direct decompositions of modules can be found in [53].

As we mentioned above, the example of the non-discrete spectral categories shows that the categories satisfying the equivalent conditions of Theorem 1 are not necessarily locally noetherian. The Theorem 1 suggests the introduction and study of a continuous variant of locally noetherian categories. However, several problems remain unsolved. It is for example not known (except in special cases) whether the condition (ii) of Theorem 1 implies that (directed) direct limits of injective objects are still injective [53]. In what follows, we will however stick to the (discrete) locally noetherian case. This case is easily isolated from the continuous case, if we require the axiom AB 6 [50] instead of axiom AB 5 (the axiom A above) as the ground axiom on our categories. We have in fact the following

THEOREM 2.- [Characterizations of locally noetherian categories.] The following conditions on an abelian category $\underline{C}$ with a set of generators are equivalent:

(i) $\underline{C}$ is locally noetherian;

(ii) [resp. (ii')] $\underline{C}$ satisfies the axiom AB 6, and every sum (resp. directed $\varinjlim$) of injectives is injective;

(iii) $\underline{C}$ satisfies AB 6 and has bounded decompositions of injectives;

(iv) $\underline{C}$ satisfies AB 6 and has a strict cogenerator;

(v) $\underline{C}$ satisfies AB 6 (AB 5 might be sufficient here), and every injective is a direct sum of indecomposable injectives;

(vi) $\underline{C}$ satisfies AB 6 and every direct limit of a direct system of essential monomorphisms is an essential monomorphism.

For indications of the proof of this theorem we refer the reader to [53] and to [50] where the condition AB 6 and its variants are discussed. This Theorem 2 contains both the Proposition 1 above as well as the theorem of [43] since a module category, and more generally a category that is locally of finite type [43], both automatically satisfy the axiom AB 6 [50].

§ 2. Locally noetherian and locally coherent categories. The conjugate category.

Gabriel has observed [24], [25] [cf. also [31], [56]], that if $\underline{C}$ is locally noetherian and $N(\underline{C})$ the category of noetherian objects of $\underline{C}$, then $N(C)$ is an abelian category, that is equivalent to a small category, and $\underline{C}$ is equivalent to the category $\mathrm{Lex}(\underline{N}(C)^{o}, \underline{Ab})$ [1] of left exact contravariant functors from $N(\underline{C})$ to $\underline{Ab}$ (= the category of abelian groups). Conversely, if $\underline{N}$ is a noetherian abelian category (i.e. an abelian category, where every object is noetherian) that is equivalent to a small category, then $\mathrm{Lex}(\underline{N}^{o}, \mathrm{Ab})$ is locally noetherian, and its category of noetherian objects is naturally equivalent to $\underline{N}$. Thus the map

$$(1) \qquad \underline{N} \longmapsto \mathrm{Lex}(\underline{N}^{o}, \underline{Ab})$$

defines a one-one correspondence between the (equivalence classes of) small abelian noetherian categories and the (equivalence classes of) locally noetherian categories.

It is natural to ask, what kind of Grothendieck categories we obtain to the right in (1), when $\underline{N}$ is replaced by an arbitrary small abelian category $\underline{D}$. To answer this, we first need the definition below. Recall first that an object C in a Grothendieck category $\underline{C}$ is said to be of finite type [43] if $\mathrm{Hom}_{\underline{C}}(C,\cdot)$ commutes with directed unions.

DEFINITION 1.- Let C be an object in a Grothendieck category $\underline{C}$. We say that C is coherent, if C is of finite type, and if for any morphism $C' \xrightarrow{f} C$, where C' is of finite type, $\mathrm{Ker}\, f$ is so too. If $\underline{C}$ has a family of generators, formed by coherent objects, then we say that $\underline{C}$ is a locally coherent category.

The following result can be found with small modifications in [22] and

---

1) If $\underline{C}$ is a category, $\underline{C}^{o}$ denotes the dual category of $\underline{C}$ [30].

its proof is essentially an adaption to the locally coherent case, of the corresponding proof of Gabriel for the special case of the locally noetherian categories.

PROPOSITION 2.- The map

$$\underline{D} \longmapsto \mathrm{Lex}(\underline{D}^{o}, \underline{Ab})$$

defines a 1-1 correspondence between the (equivalence classes of) small abelian categories and the (equivalence classes of) locally coherent Grothendieck categories. Furthermore $\underline{D}$ is naturally equivalent to the category of coherent objects of $\mathrm{Lex}(\underline{D}^{o}, \underline{Ab})$, and $\underline{D}$ is naturally embedded into $\mathrm{Lex}(\underline{D}^{o}, \underline{Ab})$, by means of the functor $D \longmapsto \mathrm{Hom}_{\underline{D}}(\cdot, D) = h_D$.

Examples of locally coherent categories:

1) Mod(.A) is locally coherent if and only if the ring A is left coherent in the sense of [9], p. 63 (cf. also [13]), i.e. if and only if every finitely generated left ideal of A is finitely presented.

2) J. Cohen has proved [15] that $\mathrm{Modgr}^{+}(A)$, where A is the Steenrod algebra, is locally coherent. This category is however not locally noetherian. [For the notation, see example 6 in the introduction.]

Now let $\underline{C}$ be a locally coherent category, and let $\mathrm{Coh}(\underline{C})$ be the category of coherent objects of $\underline{C}$. [Then, as we have seen, $\mathrm{Coh}(\underline{C})$ is abelian, equivalent to a small category, and $\underline{C}$ is equivalent to $\mathrm{Lex}(\mathrm{Coh}(\underline{C})^{o}, \underline{Ab})$.] Put $\underline{\tilde{C}} = \mathrm{Lex}(\mathrm{Coh}(\underline{C}), \underline{Ab})$. This category (which is evidently locally coherent) will be called the conjugate category of $\underline{C}$ in what follows. It is easy to see that $\underline{\tilde{C}}$ is equivalent to $\mathrm{Lex}_{\sim \varinjlim}(\underline{C}, \underline{Ab})$ = the category of left exact functors $\underline{C} \longrightarrow \underline{Ab}$ that commute with directed $\varinjlim$. Furthermore $\underline{\tilde{\tilde{C}}}$ is naturally equivalent to $\underline{C}$. Our aim in the next section is to characterize completely the categories $\underline{\tilde{C}}$, where $\underline{C}$ is a locally noetherian category. Since every object of $\mathrm{Coh}(\underline{C})^{o}$ is artinian (i.e. satisfies the descending chain condition on subobjects) if $\mathrm{Coh}(\underline{C})$ is noetherian, one might think that the conjugate categories of locally noetherian ones, would be exactly the locally coherent and locally artinian categories (locally artinian means for us: there exists a family of generators that are artinian - this is different from Oort's terminology [44]). However, these categories form only a special case of the class of conjugates of locally noetherian categories, and $\widetilde{\mathrm{Mod}(\mathbb{Z})}$ is not of this special form (cf. Example 1 following Corollary 3 of Theorem 12 in § 8). Thus we must study a descending chain condition, that is weaker than the usual one:

§ 3. Coperfect categories. The conjugate category of a locally noetherian category. Krull dimension of Grothendieck categories.

As is well known [23], [25], not only does every object C in a Grothendieck category admit a monomorphism into an injective object, but it also admits a minimal one, that is essentially unique, and that is called the injective envelope of C. The dual of this notion is the projective envelope (also called projective cover). However the existence of projective envelopes is a rare phenomenon, even if the category is a module category (so that we have sufficiently many projectives). Bass [3] called a ring R a right perfect ring if every right R-module has a projective cover, and he proved the following result:

THEOREM OF BASS. - The following conditions on a ring R are equivalent:

(i) R is right perfect;

(ii) R satisfies minimum condition on principal left ideals;

(iii) $R/\mathrm{rad}_J R$ is an artinian semi-simple ring, and $\mathrm{rad}_J R$ is right T-nilpotent [3] (rad_J = the Jacobson radical);

(iv) Every direct limit of projective right R-modules is projective;

...

Bass left as an open problem, whether (ii) is in fact equivalent to:

(ii)' R satisfies minimum condition on finitely generated left ideals;

but J.-E. BJÖRK proved recently [6] that (ii) and (ii)' are in fact equivalent. This generalizes an earlier result of S.U. CHASE (cf. Appendix to [20]) which says that (ii)' is verified for semi-primary rings (a special case of the perfect ones). In view of the BASS-BJÖRK result, it is natural to call a module M coperfect, if M satisfies the minimum condition on finitely generated submodules. More generally:

DEFINITION 2.- A Grothendieck category $\underline{C}$ is said to be (locally) coperfect if $\underline{C}$ is locally of finite type (cf. [43]), and if $\underline{C}$ has a family of generators $\{G_\alpha\}$, where each G_α is coperfect, i.e. satisfies minimum condition on finitely generated subobjects.

Remark: If $\underline{C}$ is locally coperfect, then every object of $\underline{C}$ is in fact coperfect, so we can and will omit the word "locally" in what follows.

THEOREM 3.- The map $\underline{C} \longmapsto \tilde{\underline{C}}$ (cf. § 2) defines a one-one correspondence between the (equivalence classes of) locally noetherian categories and the (equivalence classes of) locally coherent and coperfect categories.

PROOF: Suppose first that $\underline{C}$ is locally noetherian. We only have to prove that $\tilde{\underline{C}} = \text{Lex}(N(\underline{C}), \underline{Ab})$ is coperfect (cf. § 2). The set of objects $h^N = \text{Hom}_{\underline{C}}(N,\cdot)$, $N \in N(\underline{C})$ is exactly the set of coherent generators for $\tilde{\underline{C}}$, and so it is sufficient to prove that every h^N is coperfect. But since h^N is coherent, every finitely generated subobject of h^N is coherent, thus of the form $h^{N'}$. Since $N(\underline{C})^{\circ}$ is fully and faithfully embedded in $\tilde{\underline{C}}$ by $N \longmapsto h^N$, N' is a quotient of N. Thus the result follows from the fact that $N(\underline{C})$ is a noetherian category. Conversely, given a locally coherent and coperfect category $\underline{D}$, it follows that $\text{Coh}(\underline{D})$ is an artinian abelian category, and so $\tilde{\underline{D}}$ is locally noetherian.

Remark and example: A module category $\text{Mod}(.R)$ is locally coherent and coperfect if and only if R is left coherent and right perfect. This class of rings was first introduced by Chase [13], who proved that it is equivalent to say, that every product of right projective R-modules is again projective. In the commutative case, these rings coincide with the artinian ones [13], but in the general case they need not be either left or right artinian, as the following example (which was communicated to me by L.W. Small) shows: $R = \begin{pmatrix} \mathbb{Q} & \mathbb{C} \\ 0 & \mathbb{Q} \end{pmatrix}$. (For more details, and for a general theory of generalized triangular matrix rings, cf. § 8 and § 9.)

The correspondence $\underline{C} \longleftrightarrow \underline{\mathcal{T}}$ will be studied in more detail below, using the endomorphism rings of injective objects, and we will just conclude this section with a few elementary remarks, that are easy consequences of the theory of Krull dimension of Grothendieck categories [25], that we will first recall briefly:

Suppose $\underline{C}$ is a Grothendieck category. Then a full subcategory $\underline{L}$ of $\underline{C}$ is said to be a localizing subcategory [25] if $\underline{L}$ is closed under formation of subobjects, quotient objects, extensions and $\varinjlim$ (in $\underline{C}$) [(Note) In this paper every subcategory will be a strict subcategory, i.e. the subcategory should contain all isomorphic copies (in the big category) of its objects.] If $\underline{L}$ is a localizing subcategory, then it is also a Grothendieck category, the quotient category $\underline{C}/\underline{L}$ can be formed, and the natural functor $\underline{C} \xrightarrow{j^*} \underline{C}/\underline{L}$ is exact, and has a right adjoint j_* that is full and faithful [25]. Thus $\underline{C}/\underline{L}$ is also a Grothendieck category. Now let $\underline{C}_o$ be the smallest (exists!) localizing subcategory of $\underline{C}$, that contains the simple objects (S simple $\iff$ $S \neq 0$, and only 0 and S are subobjects of S) of $\underline{C}$. Form $\underline{C}/\underline{C}_o$, and let $\underline{C}_1$ be the smallest localizing subcategory of $\underline{C}$, containing those $C \in \underline{C}$, that are simple or 0 in $\underline{C}/\underline{C}_o$ etc... We get a transfinite filtration[1] $\underline{C}_o \subset \underline{C}_1 \subset \cdots \subset \cdots$ (the definition of $\underline{C}_\alpha$ for a limit ordinal is evident), and there exists an ordinal α^* such that $\underline{C}_{\alpha^*} = \underline{C}_{\alpha^*+1} = \cdots$. If $\underline{C}_{\alpha^*} = \underline{C}$, then $\underline{C}$ is said to have a Krull dimension defined, and we put $\dim \underline{C}$ = the smallest such α^*. Not every Grothendieck category has a Krull dimension. Counter-example: a continuous (non-discrete) spectral category. One can however introduce a continuous analogue of Krull dimension, and then this last category has dimension 0. However, Mod(A, $\underline{m}$) [48], [49], where A is a non-discrete archimedean valuation ring and $\underline{m}$ the maximal ideal, does not even have a Krull dimension in this more general sense.

It is easy to see that $\underline{C}$ has Krull dimension defined and equal to 0 if and only if every $C \neq 0$ contains a simple object. A locally

1) This filtration will be called the Gabriel filtration of $\underline{C}$.

noetherian category $\underline{C}$ always has a Krull dimension and this dimension is zero if and only if $\underline{C}$ is locally finite [25]. [The Grothendieck category $\underline{C}$ is said to be locally finite if $\underline{C}$ has a family of generators, formed by objects of finite length, i.e. objects that are both artinian and noetherian [25].] Finally, the Krull dimension of Mod(A), where A is a <u>commutative</u> noetherian ring, coincides with the dimension of A defined by means of chains of prime ideals (Krull) [25]. This last description is valid in some non-commutative cases too [25]. For another definition of Krull dimension (when it is $\leq \aleph_0$), see [28].

Now if $\underline{C}$ is arbitrary locally noetherian, then $\tilde{\underline{C}}$ has Krull dimension 0, for if $C \neq 0$ is an object of $\tilde{\underline{C}}$, then a minimal element in the family of finitely generated non-zero subobjects of C must be a simple object. Thus in particular $\tilde{\underline{C}}$ is locally noetherian if and only if it is locally finite, which is the case if and only if $\underline{C}$ itself is locally finite. [In the module case $\tilde{\underline{C}} = \widetilde{\text{Mod}(R)}$, considered above, we see thus that R is artinian, if and only if $\widetilde{\text{Mod}(R)}$ is locally finite, and one can prove that $\widetilde{\text{Mod}\left(\begin{pmatrix} Q & \underset{\sim}{C} \\ 0 & \underset{\sim}{Q} \end{pmatrix}\right)}$ has Krull dimension 1. More detailed results are to be found in § 8 and § 9. In particular we will see that $\widetilde{\text{Mod}(R)}$ has <u>finite</u> Krull dimension for R left coherent and right perfect, and that <u>every dimension can occur</u>.] <u>Thus</u>: The study of a locally noetherian category $\underline{C}$ of <u>arbitrary</u> Krull dimension is equivalent to the study of a category $\tilde{\underline{C}}$ of Krull dimension 0 of a special type (locally coherent and coperfect), this last category being locally finite, if and only if $\underline{C}$ is so (i.e. if and only if $\underline{C}$ is of Krull dimension 0). <u>This gives an indication of the enormous difference between the categories of Krull dimension 0 and the locally finite categories</u>.

§ 4. <u>Structure of endomorphism rings of injective objects in locally noetherian categories</u>.

In [40] Matlis proved, that if A is a commutative noetherian ring,

$\mathfrak{p}$ a prime ideal in A, and $E = E(A/\mathfrak{p})$ the injective envelope of $A/\mathfrak{p}$ in Mod(A), then the endomorphism ring of E in Mod(A) is exactly the completion of the local ring $A_{\mathfrak{p}}$. For complete (noetherian) local rings there are structure theorems, due to I.S. Cohen [18], [32]. This leads one to expect, that if I is an injective object in an arbitrary locally noetherian category $\underline{C}$, then $\mathrm{Hom}_{\underline{C}}(I,I)$ should have a natural topology and that there should be some explicit structure theory for this topological ring. We will show below, that this is indeed the case (to a certain degree) and that, furthermore, the (topological) endomorphism ring of a big injective (for a definition of that see Theorem 4 below) in $\underline{C}$, completely determines $\underline{C}$. In the special case, when the Krull dimension of $\underline{C}$ is zero (i.e. $\underline{C}$ locally finite) the above results are due to Gabriel [24], [25], who also had some indications about the general case, but as we will see later, quite new phenomena occur when we pass to the complete study of the general case.

The following well-known proposition about linearly topologized rings will be useful in what follows. Recall that a full subcategory $\underline{D}$ of a Grothendieck category $\underline{C}$ is said to be a closed subcategory of $\underline{C}$, if $\underline{D}$ is closed under formation of subobject, quotient objects and (directed) $\varinjlim$ in $\underline{C}$ (such a $\underline{D}$ is then necessarily also a Grothendieck category) and that a topology on a ring[1)]R is said to be a (left) linear topology on R, if there is a fundamental system of neighbourhoods of 0 consisting of left ideals (these ideals are then open).

PROPOSITION 3.- Let A be a ring. The map that to each (left) linear topology on A associates the full subcategory Dis(A) of Mod(A) whose objects are the discrete topological A-modules, defines a one-one correspondence between the (left) linear topologies on A, and the closed subcategories of Mod(A). Furthermore, the open left ideals $\underline{l}$ of A are exactly those left ideals such that $A/\underline{l} \in \mathrm{Dis}(A)$, and for a given left linear topology τ on A, the corresponding set of open left ideals J_τ

1) In this paper all topologies considered on rings are supposed to be compatible with the ring structure.

satisfies:

(i) $\mathfrak{a} \in J_\tau$, $\mathfrak{b} \supset \mathfrak{a} \Rightarrow \mathfrak{b} \in J_\tau$;

(ii) $\mathfrak{a}, \mathfrak{b} \in J_\tau \Rightarrow \mathfrak{a} \cap \mathfrak{b} \in J_\tau$;

(iii) $\mathfrak{a} \in J_\tau$, a arbitrary in $A \Rightarrow (\mathfrak{a}:a) = \{x \mid xa \in \mathfrak{a}\} \in J_\tau$.

Conversely, given a set J of left ideals satisfying (i)-(iii), there is a unique linear topology τ on A such that $J = J_\tau$.

(Cf. Gabriel [25], p. 411-412 and Bourbaki [9], exerc. 16, p. 157.)

Now, let $\underline{C}$ be a Grothendieck category, I an injective object of $\underline{C}$, and $A = \mathrm{Hom}_{\underline{C}}(I, I)$. It is clear that if C is an arbitrary object of $\underline{C}$, then $\mathrm{Hom}_{\underline{C}}(C, I)$ is in a natural way a left A-module [we write the composite map $C \xrightarrow{f} I \xrightarrow{\lambda} I$ as $\lambda \cdot f$]. Thus in particular, if $C \subset I$ is a subobject of I, then $l(C) = \mathrm{Hom}_{\underline{C}}(I/C, I)$ is in a natural way a left ideal of A, and the exact sequence (I is injective):

$0 \longrightarrow \mathrm{Hom}_{\underline{C}}(I/C, I) \longrightarrow \mathrm{Hom}_{\underline{C}}(I, I) \longrightarrow \mathrm{Hom}_{\underline{C}}(C, I) \longrightarrow 0$

shows that the quotient module $A/l(C)$ can be identified with $\mathrm{Hom}_{\underline{C}}(C,I)$. Consider now the case when $\underline{C}$ is locally noetherian. Then the set J of left ideals of A that contain an ideal of the form $l(N)$, where N is a noetherian subobject of I satisfies the conditions (i), (ii) and (iii) of Proposition 3 as is easily seen. [We have for example that if $N_i \subset I$, $i = 1,2$, then

$$l(N_1) \cap l(N_2) = \mathrm{Hom}_{\underline{C}}(I/(N_1+N_2), I),$$

where N_1+N_2 is the image if the natural map $N_1 \amalg N_2 \longrightarrow I$, and this image is clearly noetherian if N_i, $i = 1,2$ are so.] Thus the ring $A =$ $= \mathrm{Hom}_{\underline{C}}(I, I)$ has a natural linear topology, for which the $l(N)$ form a fundamental system of (left) ideal neighborhoods of 0. Furthermore, for this topology A is complete (by that we mean Hausdorff too). In fact, to see this we have to prove that the natural map

$$(2) \qquad A \longrightarrow \varprojlim_{N \subset I,\ N \text{ noeth.}} A/l(N)$$

is an isomorphism. But the map (2) can be identified with

$$\mathrm{Hom}_{\underline{C}}(I, I) \longrightarrow \varprojlim_{\substack{N \subset I \\ N \text{ noeth.}}} \mathrm{Hom}_{\underline{C}}(N, I)$$

and

$$\varprojlim_{N \subset I} \mathrm{Hom}_{\underline{C}}(N, I) = \mathrm{Hom}_{\underline{C}}(\varinjlim N, I) = \mathrm{Hom}_{\underline{C}}(I, I)$$

where the last "equality" follows from the fact that $\underline{C}$ is locally noetherian. Thus A is a (left) linearly topologized complete ring in a natural way. However, not every such ring is obtained in this manner, for our A has more special properties. First a definition:

DEFINITION 3.- Let A be a (left) linearly topologized ring. We say that A is (left) topologically coperfect (resp. topologically coherent, resp. topologically artinian, ...) if the Grothendieck category Dis(A) is (locally) coperfect (resp. locally coherent, resp. locally artinian, ...).

Remark. - If the topology on A is discrete, then Dis(A) = Mod(A) and the topological notions of Definition 3 coincide with the corresponding usual (discrete) algebraic notions as they should. In the general linear topological case, our conditions can also be expressed in terms of ideals of A as follows:

1) A is (left) topologically coperfect if and only if A has a fundamental system of (left) ideal neighbourhoods of 0, $\{\mathfrak{a}\}$, such that every $A/\mathfrak{a}$ satisfies minimum condition on finitely generated submodules (then $A/\mathfrak{b}$ satisfies this minimum condition for all open left ideals $\mathfrak{b}$).

2) A is (left) topologically coherent, if and only if A has a fundamental system of (left) ideal neighborhoods of 0, $\check{J}$ such that for every $\mathfrak{a} \in \check{J}$, the kernel of every map $\coprod_{1}^{t} A/\mathfrak{b}_i \longrightarrow A/\mathfrak{a}$ ($\mathfrak{b}_i$ open ideals of A) is of finite type. (It suffices even to suppose the $\mathfrak{b}_i$ equal and belonging to $\check{J}$.).

3) A is left topologically artinian, if and only if for every open left ideal $\mathfrak{a}$ of A, $A/\mathfrak{a}$ is an artinian A-module... .

PROPOSITION 4.- Let $\underline{C}$ be a locally noetherian category, I an injective of $\underline{C}$, and $A = \mathrm{Hom}_{\underline{C}}(I, I)$ with its natural linear topology. Then A is (complete and) topologically coperfect.

PROOF: Since the $A/l(N)$ ($N \subset I$, N noetherian) form a system of generators for Dis(A), and since $A/l(N) = \mathrm{Hom}_{\underline{C}}(N, I)$, it is more than sufficient to prove that for any noetherian object N in $\underline{C}$, $\mathrm{Hom}_{\underline{C}}(N, I)$ is a coperfect left A-module. But if $Af_1 + \ldots + Af_n \subset \mathrm{Hom}_{\underline{C}}(N, I)$ is a finitely generated left A-submodule ($f_i \in \mathrm{Hom}_{\underline{C}}(N, I)$), then it is easy to see that

$$Af_1 + \ldots + Af_n = \mathrm{Hom}_{\underline{C}}(N/\mathrm{Ker}(N \xrightarrow{(f_i)} \coprod_1^n I), I)$$

and from this the result follows, since N is noetherian.

Remark.- We do not know if conversely every linearly topologized, topologically coperfect, complete ring can be obtained in this way. [We would also like to know whether every right perfect ring can be obtained as an endomorphism ring of a noetherian injective object in an abelian category.] It is however well-known that the more restricted class of linearly topologized, complete, topologically artinian rings [these rings are also called strictly linearly compact rings ([10], p. 111), or i.e. S.l.k. rings [38], [39] or Leptin rings [24]] can be obtained in this manner [24], [25] (cf. also § 5). The endomorphism ring of $\underline{Q} \amalg \underline{Q}/\underline{Z}$ in $\mathrm{Mod}(\underline{Z})$ is however not topologically artinian (cf. Example 1 following Corollary 3 of Theorem 12 in § 8).

However, if I is a big injective (see Theorem 4 below) then the endomorphism ring can be characterized completely:

THEOREM 4.- Let $\underline{C}$ be a locally noetherian category, I a big injective of $\underline{C}$ (i.e. an injective that contains at least one indecomposable injective of each type), and A the endomorphism ring of I with its natural topology. Then A is complete, topologically coherent and topologically coperfect, and conversely every such ring can be obtained in this manner [1].

PROOF (First part): We know already that the ring A in Theorem 4 is complete and topologically coperfect. We prove here that A is also topologically coherent. (The last assertion of Theorem 4 will follow from Corollary 1 of Theorem 6 in § 5.)

Let us prove more generally, that if N is any noetherian object of $\underline{C}$, then the discrete left A-module $\mathrm{Hom}_{\underline{C}}(N, I)$ is coherent (I big injective!) as an object of $\mathrm{Dis}(A)$. First of all, it is clear that $\mathrm{Hom}_{\underline{C}}(N, I)$ is finitely generated as a left A-module, for if $I(N)$ is the injective envelope of N, then since N is noetherian, only a finite number of indecomposables occur in the decomposition of $I(N)$. Thus there exists an integer ν such that $I(N)$ is a direct factor of $\coprod_1^{\nu} I$, and so $\mathrm{Hom}_{\underline{C}}(N, I)$ is a quotient of $\coprod_1^{\nu} A$. Now let $T \xrightarrow{f} \mathrm{Hom}_{\underline{C}}(N, I)$ be any map from a finitely ~~generated~~ topological, discrete left A-module T to $\mathrm{Hom}_{\underline{C}}(N, I)$. We have to prove that $\mathrm{Ker}\, f$ is finitely generated. Since the $A/l(N')$ ($N' \subset I$, N' noetherian) form a set of generators for $\mathrm{Dis}(A)$, and since T is finitely generated, it must be a quotient of a finite sum of such objects, thus a quotient of an object of the form $\mathrm{Hom}_{\underline{C}}(N_1, I)$, where N_1 is noetherian, and by composition with f we get a map of A-modules:

$$\mathrm{Hom}_{\underline{C}}(N_1, I) \xrightarrow{\tilde{f}} \mathrm{Hom}_{\underline{C}}(N, I) .$$

It is sufficient to prove that $\mathrm{Ker}\, \tilde{f}$ is finitely generated. Now observe (Lemma 1 below) that since I is a big injective, there exists a unique map $N \xrightarrow{\tilde{F}} N_1$ such that $\mathrm{Hom}_{\underline{C}}(\tilde{F}, I) = \tilde{f}$. Thus $\mathrm{Ker}\, \tilde{f} = \mathrm{Hom}_{\underline{C}}(N_1/\mathrm{Im}\, \tilde{F}, I)$ and $N_1/\mathrm{Im}\, \tilde{F}$ is noetherian, and so $\mathrm{Ker}\, \tilde{f}$ must be finitely generated as we have just seen above. Thus $\mathrm{Hom}_{\underline{C}}(N, I)$ is coherent and so in

[1] We will see in § 5 that the equivalence class of $\underline{C}$ is uniquely determined by A.

particular A is topologically coherent, and therefore the first part of Theorem 4 is proved modulo the following lemma:

LEMMA 1.- *Let* $\underline{C}$ *be a locally noetherian category,* I *a big injective in* $\underline{C}$, $A = \mathrm{Hom}_{\underline{C}}(I, I)$, *and* $\mathrm{Hom}_{\underline{C}}(N_1, I) \xrightarrow{g} \mathrm{Hom}_{\underline{C}}(N_2, I)$ *a map of left A-modules, where* N_1 and N_2 *are noetherian. Then there is a* (necessarily) *unique morphism* $N_2 \xrightarrow{G} N_1$ *such that* $\mathrm{Hom}_{\underline{C}}(G,I) = g$.

PROOF: Let $\coprod_{s_1} I$ and $\coprod_{s_2} I$ be injectives containing N_1 and N_2. (Here we can suppose that s_1 and s_2 *are finite sets*, cf. proof of Theorem 4 above.) Thus we have two exact sequences:

$$0 \longrightarrow N_i \longrightarrow \coprod_{s_i} I \longrightarrow K_i \longrightarrow 0 \quad (i = 1,2).$$

Take $\mathrm{Hom}_{\underline{C}}(\cdot, I)$ of these two sequences, and consider the diagram:

$$(3) \qquad \begin{array}{ccccccccc} 0 & \longrightarrow & \mathrm{Hom}_{\underline{C}}(K_1, I) & \longrightarrow & \mathrm{Hom}_{\underline{C}}(\coprod_{s_1} I, I) & \longrightarrow & \mathrm{Hom}_{\underline{C}}(N_1, I) & \longrightarrow & 0 \\ & & & & \Big\downarrow \varphi & & \Big\downarrow g & & \\ 0 & \longrightarrow & \mathrm{Hom}_{\underline{C}}(K_2, I) & \longrightarrow & \mathrm{Hom}_{\underline{C}}(\coprod_{s_2} I, I) & \longrightarrow & \mathrm{Hom}_{\underline{C}}(N_2, I) & \longrightarrow & 0 \end{array}$$

Since $\mathrm{Hom}_{\underline{C}}(\coprod_{s_1} I, I) = \coprod_{s_1} A$ is projective, there exists a map of A-modules φ making the right square of (3) commute, and this map is necessarily induced by a map $\coprod_{s_2} I \xrightarrow{\Phi} \coprod_{s_1} I$. It is now easy to see, using (3) and the fact that I is a cogenerator, that the composed zig-zag map from N_2 to K_1 in the diagram

$$(4) \qquad \begin{array}{ccccccccc} 0 & \longrightarrow & N_1 & \longrightarrow & \coprod_{s_1} I & \longrightarrow & K_1 & \longrightarrow & 0 \\ & & & & \Big\uparrow \Phi & & & & \\ 0 & \longrightarrow & N_2 & \longrightarrow & \coprod_{s_2} I & \longrightarrow & K_2 & \longrightarrow & 0 \end{array}$$

is zero. Thus from (4) we get an induced map $N_2 \xrightarrow{G} N_1$, and one verifies immediately that $\mathrm{Hom}_{\underline{C}}(G, I) = g$. Uniqueness of G follows from the fact that I is a cogenerator, and so in particular the lemma is proved, and thus also the first part of Theorem 4.

§ 5.- Explicit realization of the dual of a locally noetherian category.

Recall that if τ is an arbitrary small (or equivalent to a small) category, then the category of pro-objects of τ, denoted by $\mathrm{Pro}(\tau)$, is defined as follows (cf. [31], [56]):

The objects of $\mathrm{Pro}(\tau)$ are the inverse systems $\{T_\alpha\}_{\alpha \in I}$, with I a directed set, and the morphisms are given by the formula:

$$\mathrm{Hom}_{\mathrm{Pro}(\tau)}(\{T_\alpha\}_{\alpha \in I}, \{T'_\beta\}_{\beta \in J}) = \varprojlim_{\beta} \varinjlim_{\alpha} \mathrm{Hom}_\tau(T_\alpha, T'_\beta)$$

A pro-object $\{T_\alpha\}$ is called strict [31] if all transition morphisms $T_{\alpha'} \longrightarrow T_\alpha$ $(\alpha' > \alpha)$ are epimorphisms.

It follows in particular from [17], that if τ is an abelian small (or equivalent to a small) category, then $\mathrm{Pro}(\tau)$ is naturally equivalent to $\mathrm{Lex}(\tau, \underline{\mathrm{Ab}})^\circ$ (cf. § 2). Furthermore, if τ is artinian (abelian) then every pro-object of τ is isomorphic to a strict pro-object [24], [25], [31], [44], [56]. We will now prove that if τ (abelian and artinian) can be realized as a full subcategory of a module category Mod(R), stable under the formation of kernels and cokernels (thus also under the formation of Im, Coim ...) in Mod(R) [1)], then the category $\mathrm{Lex}(\tau, \underline{\mathrm{Ab}})^\circ = \mathrm{Pro}(\tau)$ can be given a much more precise realization as a category of linear-topological R-modules. Examples for the possibility of such realizations [which go back to the topological duality theory for abelian groups and vector spaces (see notably p. 79-80 of [36] and [35])] can be found in [24], [25], [31], [35], [40], [56] [2)].

THEOREM 5.- Let R be a ring, τ an abelian artinian category, that is a full subcategory of Mod(R), that is stable under the formation of

---

1) We will see later that such a realization can always be constructed, and this even in a canonical way. But there are also other interesting cases, besides the canonical one ...

2) In this paper we restrict ourselves to the abelian case, but there should evidently also exist a corresponding theory of locally noetherian toposes [60].

kernels and cokernels in Mod(R). Then the dual of the locally noetherian category Lex(τ, Ab) is naturally equivalent to the category $Mod_\tau(R)$, whose objects are the linearly topologized, complete and separated R-modules M (R considered as discrete) that have a fundamental system of neighbourhoods of 0, formed by submodules M_α such that $M/M_\alpha \in \tau$, and whose morphisms are the R-linear continuous maps. Furthermore, the kernels cokernels, images, coimages, products, $\varprojlim$ in $Mod_\tau(R)$ are the algebraic kernels, cokernels, ... equipped with the induced, quotient, ..., product, ... topology. [1)]

Caution: When τ is realized as in Theorem 5, then the objects of τ, although artinian in τ, are not necessarily so in Mod(R) as we will clearly see in the examples considered later. Thus we have to be careful in the proofs below.

START OF THE PROOF OF THEOREM 5: We will first study the category $Mod_\tau(R)$ in detail, and notably prove the last assertion about kernels, cokernels, ..., $\varprojlim$. The proof of this is long, and is based upon eight propositions, some of them of independent interest. The rest of the theorem will then follow almost automatically from these propositions.

PROPOSITION 5.- Let R be a ring, τ a full abelian subcategory of Mod(R), stable under the formation of kernels and cokernels (not necessarily artinian!), and let M be a τ-linearly topologized R-module (defined as in Theorem 5, but no completeness or Hausdorff condition is required. Then an open submodule U of M is τ-linearly topologized for the induced topology if and only if $M/U \in \tau$.

PROOF: If $M/U \in \tau$, then U is τ-linearly topologized for the induced topology. In fact: the induced topology on U is defined by the open submodules of M, contained in U. If V is such a submodule, then since M is τ-linearly topologized, V contains an M- (thus U-) open

1) The $\varinjlim$ in $Mod_\tau(R)$ are however much more complicated.

submodule V' such that $M/V' \in \tau$. Thus we have an exact sequence in $\mathrm{Mod}(R)$

$$0 \longrightarrow U/V' \longrightarrow M/V' \longrightarrow M/U \longrightarrow 0,$$

where M/V' and M/U are in τ. Thus U/V' is also in τ, since τ is closed under the formation of kernels in $\mathrm{Mod}(R)$.

Conversely, suppose that U is a τ-linearly topologized open submodule of M. Then U contains an open V such that $M/V \in \tau$ (M is τ-linearly topologized) and V contains an open V' such that $U/V' \in \tau$ (U is τ-linearly topologized). The natural exact sequence

$$U/V' \longrightarrow M/V \longrightarrow M/U \longrightarrow 0$$

shows that M/U is in τ, for τ is closed under the formation of cokernels in $\mathrm{Mod}(R)$ and so the Proposition 5 is proved.

COROLLARY.- *Let* M *be as in the Proposition* 5. *Then the intersection and sum of two open* τ-*linearly topologized submodules of* M *is again of the same type*.

PROOF. Let U and V be the submodules in question. We have an exact sequence

$$(5) \qquad 0 \longrightarrow U/U \cap V \longrightarrow M/V \longrightarrow M/(U+V) \longrightarrow 0$$

Here $U \cap V$ is an open submodule of U and U that is τ-linearly topologized. Thus $U/U \cap V$ is a quotient of an object in τ and a subobject of $M/V \in \tau$. Thus since τ is closed under images in $\mathrm{Mod}(R)$, we obtain that $U/U \cap V$ is in τ, i.e. $U \cap V$ is a τ-linearly topologized module. Finally, the exact sequence (5) gives that $M/(U+V) \in \tau$, and since $U+V$ is open, it is τ-linearly topologized too.

PROPOSITION 6.- *Let* τ *be as in Proposition* 5, *and let* M_1 *and* M_2 *be two* τ-*linearly topologized* R-*modules, and* $M_1 \xrightarrow{f} M_2$ *a continuous* R-*linear map*. *Suppose also that* M_2 *is Hausdorff*. *Then* $\mathrm{Ker}\, f$ *is the* (directed) *intersection of those open submodules of* M_1, *that are*

τ-linearly topologized and that contain Ker f.

PROOF: Let $\{M_{2,i}\}_{i\in I}$ be the (directed) set of all τ-linearly topologized open R-submodules of M_2, and consider the commutative exact diagram:

$$(6)\qquad \begin{array}{ccccccccc} & & & & 0 & & & & 0 \\ & & & & \uparrow & & & & \uparrow \\ & & & & M_1/f^{-1}(M_{2,i}) & & \hookrightarrow & & M_2/M_{2,i} \\ & & & & \uparrow & & & & \uparrow \\ 0 & \longrightarrow & \operatorname{Ker} f & \longrightarrow & M_1 & & \xrightarrow{\ f\ } & & M_2 \\ & & \| & & \uparrow & & & & \uparrow \\ 0 & \longrightarrow & \operatorname{Ker} f & \longrightarrow & f^{-1}(M_{2,i}) & & \longrightarrow & & M_{2,i} \\ & & & & \uparrow & & & & \uparrow \\ & & & & 0 & & & & 0 \end{array}$$

It follows that the first line is a monomorphism, and since $f^{-1}(M_{2,i})$ is open, it contains an open submodule V such that $M_1/V \in \tau$. Thus we have two maps

$$M_1/V \xrightarrow{\text{onto}} M_1/f^{-1}(M_{2,i}) \xrightarrow{\text{mono}} M_2/M_{2,i}$$

where the two extreme objects are in τ. It follows that the middle object is in τ, and so $f^{-1}(M_{2,i})$ is τ-linearly topologized. Now we pass to the inverse limit of the last row of (6), and use that M_2 is Hausdorff. This gives the exact sequence:

$$0 \longrightarrow \operatorname{Ker} f \longrightarrow \bigcap_i f^{-1}(M_{2,i}) \longrightarrow 0$$

and so the Proposition 6 is proved.

From now on we have to suppose that τ is artinian too, and our results depend heavily on this assumption.

PROPOSITION 7.- With the notations and the hypotheses of Theorem 5, let

$$(7)\qquad 0 \longrightarrow (C_\alpha) \longrightarrow (D_\alpha) \xrightarrow{(f_\alpha)} (E_\alpha) \longrightarrow 0$$

be an exact sequence of directed inverse systems in Mod(R). Suppose that each C_α is in τ. Then the sequence of R-modules

$$0 \longrightarrow \varprojlim C_\alpha \longrightarrow \varprojlim D_\alpha \longrightarrow \varprojlim E_\alpha \longrightarrow 0,$$

<u>obtained from</u> (7) <u>by applying</u> $\varprojlim$, <u>is exact</u>.

PROOF: The only thing to verify is that $\varprojlim D_\alpha \longrightarrow \varprojlim E_\alpha$ is onto. So let $\{\xi_\alpha\}$ be an element of $\varprojlim E_\alpha$, and consider the sets $X_\alpha = f_\alpha^{-1}(\xi_\alpha)$. We have $X_\alpha = x_\alpha + C_\alpha$, where $f_\alpha(x_\alpha) = \xi_\alpha$. It is clear that if $D_\beta \xrightarrow{q_{\alpha\beta}} D_\alpha$ $(\beta > \alpha)$ are the transition maps of (D_α), then $q_{\alpha\beta}(X_\beta) \subset X_\alpha$. Let $\check{q}_{\alpha\beta} : X_\beta \longrightarrow X_\alpha$ be the restriction of $q_{\alpha\beta}$. Then $(X_\alpha, \check{q}_{\alpha\beta})$ is an inverse system of sets, and it is even an inverse system of affine varieties. Everything will follow, if we can prove that $\varprojlim X_\alpha$ is non-empty. For this we will use Bourbaki [8], Chap. 1, theorème 1, p. 138. We will choose as $\mathfrak{S}_\alpha$ (cf. <u>loc. cit.</u>) the empty set, <u>and</u> those affine subvarieties of X_α, whose associated linear variety is in τ. Let us prove that the conditions (i)-(iv) of <u>loc. cit.</u> are verified. Observe first, that the associated linear map $\breve{q}_{\alpha\beta}$ of $\check{q}_{\alpha\beta}$ is the transition morphism $C_\beta \longrightarrow C_\alpha$ of (C_α), thus a map between objects of τ. To verify (i), it is thus sufficient to observe that the intersection in Mod(R) of two subobjects in τ of C_α is still in τ, and then use the artinian hypothesis on τ. The condition (ii) follows in the same way. Let us prove (iii). Let $\eta_\alpha \in X_\alpha = x_\alpha + C_\alpha$. Then either $\check{q}_{\alpha\beta}^{-1}(\eta_\alpha)$ is empty (thus in $\mathfrak{S}_\alpha$), <u>or</u> we have an element η with $\check{q}_{\alpha\beta}(\eta) = \eta_\alpha$. But then $\check{q}_{\alpha\beta}^{-1}(\eta_\alpha) = \eta + \mathrm{Ker}(\breve{q}_{\alpha\beta})$, and $\mathrm{Ker}\,\breve{q}_{\alpha\beta}$ is in τ, since $\breve{q}_{\alpha\beta}$ is a map between objects of τ. Thus (iii) is verified. As to (iv), it is verified in the same way; one uses the fact that the image in Mod(R) of a map between objects of τ is still in τ. Thus we can apply theorème 1, <u>loc.cit.</u>, and the Proposition 7 follows.

PROBLEM: It is easy to see that Proposition 7 can be formulated as saying that $\varprojlim^{(1)} C_\alpha = 0$ [46]. Is it true that $\varprojlim^{(i)} C_\alpha = 0$, $i \geq 1$, under the same hypotheses? [This is true at least in some special cases [47].]

Now that we have Proposition 7 we can easily continue to develop the theory of $Mod_\tau(R)$. In all that follows, we suppose that we are under the hypotheses of Theorem 5 (last time we recall it).

PROPOSITION 8.- Let M be an object of $Mod_\tau(R)$, M_i a filtered decreasing family of open τ-linearly topologized submodules (they are then in $Mod_\tau(R)$). Then the sequence of R-modules

$$0 \longrightarrow \bigcap_i M_i \longrightarrow M \longrightarrow \varprojlim_i M/M_i \longrightarrow 0$$

is exact [1].

PROOF: The proof is analogous to that of prop. 10, p. 391 of Gabriel [25], but we have to make τ appear explicitly everywhere. Let $\{U_\alpha\}$ be the directed (cf. the Corollary of Proposition 5) family of open τ-linearly topologized submodules of M. Consider the commutative exact diagram

$$(8)\qquad \begin{array}{ccccccc} 0 \longrightarrow & \bigcap_\alpha U_\alpha & \longrightarrow & M & \longrightarrow & \varprojlim M/U_\alpha & \\ & \downarrow & & \searrow & & \downarrow & \\ 0 \longrightarrow & \varprojlim_{i,\alpha}(M_i+U_\alpha)/M_i & \longrightarrow & \varprojlim_i M/M_i & \longrightarrow & \varprojlim_{i,\alpha} M/(M_i+U_\alpha) & \end{array}$$

Here $(M_i+U_\alpha)/M_i \xrightarrow{\sim} U_\alpha/M_i \cap U_\alpha$ and by Proposition 5 and its Corollary this object is in τ, and so by Proposition 7, the last morphism of the last line of (8) is onto, and in the same way we see that the vertical morphism to the right of (8) is onto. Since $M \longrightarrow \varprojlim M/U_\alpha$ is onto (it is even an isomorphism), the Proposition 8 now follows from the diagram (8) if we can show that $\varprojlim_{i,\alpha}(M_i+U_\alpha)/M_i = 0$. For this it is sufficient to prove that $\varprojlim_\alpha(M_i+U_\alpha)/M_i = 0$, and this is clear since there is an α such that $U_\alpha = M_i$, and so the proposition is proved.

---

1) $\bigcap$ and $\varprojlim$ are here taken in $Mod(R)$, but we will see later that they coincide with the corresponding $\bigcap$ and $\varprojlim$ in $Mod_\tau(R)$.

Remark: Proposition 8 as well as the following one, will be made more precise below (the axiom AB 5^* [30] is valid in $\mathrm{Mod}_\tau(R)$ ...).

PROPOSITION 9. - *Let* $M \in \mathrm{Mod}_\tau(R)$ *and let* U *be a* τ-*linearly topologized open submodule and* $\{M_i\}$ *a directed decreasing family of* τ-*linearly topologized open submodules of* M. *Then*

$$U + \bigcap_i M_i = \bigcap_i (U + M_i).$$

PROOF (à la Gabriel): Consider the exact commutative diagram

$$\begin{array}{ccccccccc}
0 & \longrightarrow & U & \longrightarrow & M & \longrightarrow & M/U & \longrightarrow & 0 \\
 & & \downarrow l & & \downarrow j & & \downarrow h & & \\
0 & \longrightarrow & \varprojlim U/U\cap M_i & \longrightarrow & \varprojlim M/M_i & \longrightarrow & \varprojlim M/(M_i+U) & \dashrightarrow & 0
\end{array}$$

Since U is in $\mathrm{Mod}_\tau(R)$ for the induced topology, and since $U/U\cap M_i$ is in τ (Corollary to Proposition 5) we get by Proposition 8 that l is onto. (The same proposition shows that j is onto.) Thus the snake lemma ([9], Chap. 1, § 1, n° 4) shows that $\mathrm{Ker}\, h = \mathrm{Im}(\mathrm{Ker}\, j)$. But $\mathrm{Ker}\, h = \bigcap_i [(M_i+U)/U] = \dfrac{\bigcap_i (M_i+U)}{U}$, $\mathrm{Ker}\, j = \bigcap_i M_i$, and the image of $\mathrm{Ker}\, j$ in $\mathrm{Ker}\, h$ is thus $\dfrac{\bigcap_i M_i + U}{U}$ and so the proposition follows.

PROPOSITION 10.- *Let* $M_1 \xrightarrow{f} M_2$ *be a morphism in* $\mathrm{Mod}_\tau(R)$. *Then the kernel of* f *in* $\mathrm{Mod}(R)$, *equipped with the induced topology, belongs to* $\mathrm{Mod}_\tau(R)$ [so this is also the kernel of f in $\mathrm{Mod}_\tau(R)$].

PROOF: By Proposition 6 we have

$$(9) \qquad \mathrm{Ker}\, f = \bigcap_{\substack{\mathrm{Ker} f \subset M_{1,i} \subset M_1 \\ M_{1,i} \text{ open},\ M_1/M_{1,i} \in \tau}} M_{1,i} \qquad \text{(directed intersection)}$$

To prove that $\mathrm{Ker}\, f$ with the induced topology is in $\mathrm{Mod}_\tau(R)$, it is sufficient to prove that every open submodule $U \subset M_1$ contains a smaller open submodule V such that $\mathrm{Ker}\, f/V \cap \mathrm{Ker}\, f$ is in τ. (The completeness of Ker f for the induced topology is trivial.). We may suppose that U is such that $M_1/U \in \tau$, and we will prove that then we can in fact take $V = U$. Consider the diagram

$$\begin{array}{ccc} \bigcap_i M_{1,i} = \text{Ker } f & \hookrightarrow & M_1 \\ & & \downarrow g \\ & & M_1/U \\ & & \downarrow \\ & & 0 \end{array}$$

It is clear that $g(M_{1,i})$ is in τ, for $g(M_{1,i}) = \dfrac{M_{1,i}}{U \cap M_{1,i}}$ and we can use Proposition 5 and its Corollary.

Thus the $g(M_{1,i})$ form a filtered decreasing set of subobjects in τ of $M_1/U \in \tau$. Since τ is artinian, there is a minimum element of this family, say $g(M_{1,i^*})$. Thus

(10) $$M_{1,i^*} + U = M_{1,i} + U, \quad i \geq i^*.$$

But according to Proposition 9 we have

$$\bigcap_{i \geq i^*} (M_{1,i}+U) = \bigcap_{i \geq i^*} (M_{1,i}) + U$$

and this together with (10) and (9) gives

$$M_{1,i^*} + U = \text{Ker } f + U.$$

Thus $\text{Ker } f/U \cap \text{Ker } f = (\text{Ker } f + U)/U = (M_{1,i^*} + U)/U = g(M_{1,i^*})$ belongs to τ and the proposition is proved.

COROLLARY.- Under the hypothesis of Proposition 10, the quotient in Mod(R), $M_1/\text{Ker } f$, equipped with the quotient topology, belongs to $\text{Mod}_\tau(R)$.

PROOF: Let U be an open submodule of M_1 such that $M_1/U \in \tau$. Then by the proof of Proposition 10 we know that

$(\text{Ker } f + U)/U \xrightarrow{\sim} \text{Ker } f/V \cap \text{Ker } f$ belongs to τ, and so the exact sequence

(11) $0 \longrightarrow \operatorname{Ker} f/V \cap \operatorname{Ker} f \longrightarrow M_1/U \longrightarrow M_1/(\operatorname{Ker} f + U) \longrightarrow 0$

shows that $M_1/(\operatorname{Ker} f + U) \in \tau$, and so $M_1/\operatorname{Ker} f$ is τ-linearly topologized for the quotient topology. If we pass to the inverse limit in (11), and use Proposition 7, and the fact that M_1 and Ker f are complete [1], we obtain that $M_1/\operatorname{Ker} f$ is also complete for the quotient topology, thus an object of $\operatorname{Mod}_\tau(R)$, and the corollary follows.

Now we wish to compare $M_1/\operatorname{Ker} f$ with its quotient topology, and Im f [image in Mod(R)] with the induced topology from M_2.

PROPOSITION 11.- <u>With the notations and hypothesis of Proposition</u> 10, <u>the natural algebraic isomorphism</u> $M_1/\operatorname{Ker} f \xrightarrow{h} \operatorname{Im} f$ <u>is a topological isomorphism, when</u> $M_1/\operatorname{Ker} f$ <u>and</u> Im f <u>are given the quotient and the induced topology, respectively</u>.

PROOF: Since the algebraic isomorphism h is induced by a continuous map, it is clearly continuous. Thus it is sufficient to prove that every open submodule of $M_1/\operatorname{Ker} f$ is also open in Im f (we will identify $M_1/\operatorname{Im} f$ and Im f algebraically). Now let (V_i) be the set of open submodules of M_2, such that $M_2/V_i \in \tau$, and consider $(V_i \cap \operatorname{Im} f)$ (this is a fundamental system of neighbourhoods of 0 for the induced topology on Im f). The $V_i \cap \operatorname{Im} f$ are open for the induced topology on Im f, thus open for the quotient topology, and so $\operatorname{Im} f/V_i \cap \operatorname{Im} f$ is the quotient of an object of τ. Since also $\operatorname{Im} f/V_i \cap \operatorname{Im} f =$
$= (\operatorname{Im} f + V_i)/V_i$ is a subobject of $M_2/V_i \in \tau$, we obtain that $\operatorname{Im} f/V_i \cap \operatorname{Im} f \in \tau$.

Now let P be an open submodule of Im f <u>for the quotient topology</u>, such that $\operatorname{Im} f/P \in \tau$ [recall that by Corollary of Proposition 10, Im f with the <u>quotient topology</u> belongs to $\operatorname{Mod}_\tau(R)$], and consider the

---

[1] The completeness of Ker f follows from Proposition 10.

diagram

$$\begin{array}{ccccc} 0 \longrightarrow & V_i \cap \operatorname{Im} f & \longrightarrow & \operatorname{Im} f \\ & & & \downarrow k \\ & & & \operatorname{Im} f/P \\ & & & \downarrow \\ & & & 0 \end{array}$$

I claim that $k(V_i \cap \operatorname{Im} f) \in \tau$. We have

$$k(V_i \cap \operatorname{Im} f) = \frac{(V_i \cap \operatorname{Im} f) + P}{P} = \frac{V_i \cap \operatorname{Im} f}{P \cap V_i \cap \operatorname{Im} f} .$$

But $V_i \cap \operatorname{Im} f$ is open τ-linearly topologized submodule of $\operatorname{Im} f$ for the quotient topology, and since P is also such a submodule, $P \cap \operatorname{Im} f \cap V_i$ is so too (Corollary to Proposition 5) and so $(V_i \cap \operatorname{Im} f)/(P \cap V_i \cap \operatorname{Im} f) \in \tau$ by Proposition 5. Since τ is artinian, the decreasing family of objects in τ $\{k(V_i \cap \operatorname{Im} f)\}_i$ has a minimum element, say $k(V_{i^*} \cap \operatorname{Im} f)$, i.e. $(V_{i^*} \cap \operatorname{Im} f) + P = (V_i \cap \operatorname{Im} f) + P$, $i \geq i^*$.

Now apply the Proposition 9 (quotient topology). We obtain

$$(V_{i^*} \cap \operatorname{Im} f) + P = (\bigcap_{i \geq i^*} (V_i \cap \operatorname{Im} f)) + P = P$$

(the last equality follows from the fact that M_2 is Hausdorff). Thus $V_{i^*} \cap \operatorname{Im} f \subset P$. We have thus proved that P that is open for the quotient topology, is also open for the induced topology, and so these two topologies coincide and the Proposition 11 is proved.

PROPOSITION 12.- <u>Let</u> $M \in \operatorname{Mod}_\tau(R)$. <u>In order that a</u> R-<u>submodule</u> N <u>of</u> M, <u>equipped with the induced topology</u>, <u>belongs to</u> $\operatorname{Mod}_\tau(R)$, <u>it is necessary and sufficient that</u>

(12) $$N = \bigcap_{\substack{N \subset U \subset M \\ U \text{ open, } M/U \in \tau}} U$$ (This is automatically a directed intersection, cf. Proposition 5.)

PROOF: <u>If</u> (12) is verified, then we have an exact sequence

$$0 \longrightarrow N \longrightarrow M \xrightarrow{f} \prod_U M/U = M_2 \qquad \Big[\text{(product over the U:s of (12)}\Big]$$

If we equip M_2 with the product topology (every factor is given the discrete topology), then we get an object of $\mathrm{Mod}_\tau(R)$ as is easily seen, and f is continuous. Thus $N = \mathrm{Ker}\, f$, belongs to $\mathrm{Mod}_\tau(R)$ (Proposition 10). Conversely suppose that N with the induced topology belongs to $\mathrm{Mod}_\tau(R)$, and let U be an open submodule of M with quotient in τ. Then $U \cap N$ is open in N, and thus $N/U \cap N$ is a quotient of an object of τ. But the exact sequence

(13) $\quad 0 \longrightarrow N/U \cap N \longrightarrow M/U \longrightarrow M/(U+N) \longrightarrow 0$

shows that $N/U \cap N$ is a subobject of an object of τ too, thus itself an object of τ, and so also $M/(U+N) \in \tau$ by (13).

If we pass to the $\varprojlim$ in (13), using the Proposition 7, we get an exact commutative diagram

$$\begin{array}{ccccccccc}
0 \longrightarrow & \varprojlim N/U \cap N & \longrightarrow & \varprojlim M/U & \longrightarrow & \varprojlim M/(U+N) & \longrightarrow 0 \\
& \uparrow & & \uparrow & & \uparrow & \\
0 \longrightarrow & N & \longrightarrow & M & \longrightarrow & M/N & \longrightarrow 0
\end{array}$$

where the first two vertical morphisms are isomorphisms. Thus the third one is so too, and this gives

$$N = \bigcap_{\substack{N \subset V \subset M \\ V \text{ open},\ M/V \in \tau}} V$$

and so the Proposition 12 is proved.

END OF THE PROOF OF THEOREM 5: It now follows from what we have done, that $\mathrm{Mod}_\tau(R)$ is an abelian category, where the kernels, cokernels, images etc. are the algebraic kernels, cokernels, ... equipped with the induced, quotient, ... topology. It is also easy to see that $\mathrm{Mod}_\tau(R)$ has arbitrary $\varprojlim$, and that they are the algebraic $\varprojlim$ equipped with the $\varprojlim$ topology. To show that the $\varprojlim$ over directed sets are exact functors in $\mathrm{Mod}_\tau(R)$, essentially according to the dual of proposition 1.8 in [30] [a complete proof of what we need can be found in Chapter III of Mitchell's book [41] (use proposition 1.2, lemma 1.5 and theorem 1.9 of loc.cit.)], it is sufficient to prove that if $\{M_\alpha\}$ is a directed decreasing family of $\mathrm{Mod}_\tau(R)$-subobjects of $M \in \mathrm{Mod}_\tau(R)$, then the sequence

$$0 \longrightarrow \bigcap_\alpha M_\alpha \longrightarrow M \longrightarrow \varprojlim M/M_\alpha \longrightarrow 0$$

is exact [in Mod(R) or $\mathrm{Mod}_\tau(R)$ - it is the same thing as we have just seen]. But we know (Proposition 12) that every $M_\alpha = \bigcap_{\substack{M \subset V_{\alpha j} \subset M,\ V_{\alpha j} \text{ open} \\ M/V_{\alpha j} \in \tau}} V_{\alpha j}$

Thus we get an exact sequence of inverse systems

$$0 \longrightarrow \{V_{\alpha j}\} \longrightarrow \{M\} \longrightarrow \{M/V_{\alpha j}\} \longrightarrow 0.$$

But the set of these $V_{\alpha j}$ (α and j vary) ordered by inclusion is <u>directed</u>. Thus Proposition 8 gives an exact sequence

$$(14) \qquad 0 \longrightarrow \bigcap V_{\alpha j} \longrightarrow M \longrightarrow \varprojlim M/V_{\alpha j} \longrightarrow 0$$

and by the same proposition and the associativity of $\varprojlim$ we get from (14) if we first pass to the limit over each j that

$$0 \longrightarrow \bigcap_\alpha M_\alpha \longrightarrow M \longrightarrow \varprojlim M/M_\alpha \longrightarrow 0$$

is exact, and so we have verified the axiom AB 5^* for $\mathrm{Mod}_\tau(R)$.

It is clear that the objects of τ, considered as modules with the discrete topology, are the artinian objects in $\mathrm{Mod}_\tau(R)$ and τ is fully embedded in $\mathrm{Mod}_\tau(R)$. Furthermore, they form a <u>family</u> [τ is (equivalent to) a small category] of cogenerators for $\mathrm{Mod}_\tau(R)$. In fact, let M be an object of $\mathrm{Mod}_\tau(R)$, and let $\{M_\alpha\}$ be the open submodules with quotient M/M_α in τ. By hypothesis (M Hausdorff) we have the exact sequence

$$0 \longrightarrow M \longrightarrow \prod M/M_\alpha$$

and this proves the assertion.

We have thus proved that the dual category of $\mathrm{Mod}_\tau(R)$ is a locally noetherian category.

Finally, consider the functor

$$F : \mathrm{Pro}(\tau) \longrightarrow \mathrm{Mod}_\tau(R)$$

that is defined on objects by:

$$\{C_\alpha\} \longmapsto \varprojlim C_\alpha \quad \text{(with the } \varprojlim \text{ topology)}$$

and that is given on morphisms in the natural way.

<u>This functor is an equivalence of categories</u>: Since every object of $\mathrm{Mod}_\tau(R)$ is clearly isomorphic to an object of the form $F(\{C_\alpha\})$ (where $\{C_\alpha\}$ is even a <u>strict</u> pro-object), it is sufficient to prove that the natural map

$$\varprojlim_\beta \varinjlim_\alpha \mathrm{Hom}_\tau(C_\alpha, C'_\beta) \longrightarrow \mathrm{Hom}_{\mathrm{Mod}_\tau(R)}(\varprojlim C_\alpha, \varprojlim C'_\beta)$$

is an isomorphism. But for this it is sufficient to prove that for any $T \in \tau$, the natural map

$$(15) \qquad \varinjlim_\alpha \mathrm{Hom}_\tau(C_\alpha, T) \longrightarrow \mathrm{Hom}_{\mathrm{Mod}_\tau(R)}(\varprojlim C_\alpha, T)$$

is an isomorphism. But since we now know that the dual of $\mathrm{Mod}_\tau(R)$ is locally noetherian, the assertion that (15) is an isomorphism, follows from the dual of the last part of corollary 1, p. 358 of [25]. Thus the Theorem 5 is completely proved.

We now pass to some applications of Theorem 5. The most interesting case for us here will be the case when $\underline{C}$ is a locally noetherian category, R the endomorphism ring of a big injective I in $\underline{C}$ and where $\tau \subset \mathrm{Mod}(.R)$ will now be defined. We know by Lemma 1 of § 4 that the map

(16) $N(\underline{C})^{\circ} \ni C \longmapsto Hom_{\underline{C}}(C, I) \in Mod(R)$ ($N(\underline{C})$ = the category of noetherian objects of of $\underline{C}$)

is a full embedding. Furthermore R has a natural topology, and it is easy to see that (16) defines a natural equivalence between $N(\underline{C})^{\circ}$ and the category Coh(Dis(R)) of coherent objects in Dis(R). Now if we take τ = Coh(Dis(R)) $\left[\xrightarrow{\sim} N(\underline{C})^{\circ}\right]$, then all the conditions of Theorem 5 are easily seen to be verified, and so $\underline{C}^{\circ} \xrightarrow{\sim} Lex(N(\underline{C})^{\circ}, Ab)^{\circ}$ is naturally equivalent to $Mod_{\tau}(R)$. However, due to the special form of τ, this last category can be interpreted as the category of topologically coherent, complete R-modules, denoted by TC(R), whose objects are the linearly topologized R-modules M, that are complete, and that have a fundamental system of open neighbourhoods of 0, formed by submodules M_{α} such that M/M_{α} is coherent considered as an object of Dis(R) and whose morphisms are the continuous linear map. _Furthermore_ R _with its natural topology_ is an object of TC(R), considered as a left R-module, and the objects of TC(R) are _automatically_ topological left R-modules, when R is given this topology that is coarser than the discrete one. Finally every $Hom_{\underline{C}}(C, I)$ with its natural linear topology [defined by the $Hom_{\underline{C}}(C/N, I)$ ($N \subset C$, noetherian)], belongs to TC(R) and the equivalence $\underline{C}^{\circ} \xrightarrow{\sim} TC(R)$ is given by $C \longmapsto Hom_{\underline{C}}(C, I)$.

In the reasoning above, it is not necessary to start with $\underline{C}$. In fact, if R is an arbitrary left topologically coherent and topologically coperfect ring (R not necessarily complete), then Coh(Dis(R)) is an artinean abelian category, and so $TC(R)^{\circ} \xrightarrow{\sim} Lex(Coh(Dis(R), \underline{Ab})$ is locally noetherian although R is not in TC(R) in this case, and its conjugate category $Lex(Coh(Dis(R))^{\circ}, \underline{Ab})$ is equivalent to Dis(R). Let us summarize and complete the results obtained:

THEOREM 6.- _Let_ $\underline{C}$ _be a locally noetherian category_, I _a big injective in_ $\underline{C}$, $R = Hom_{\underline{C}}(I, I)$ _the endomorphism ring of_ I _with its natural_ linear topology [1], and TC(R) the category of topologically coherent complete linear-topological R-modules (they are then automatically topologically coperfect). Then we have a natural functor

$$\underline{C}^{o} \ni C \longmapsto \mathrm{Hom}_{\underline{C}}(C, I) \in TC(R) \quad \text{(natural topology on } \mathrm{Hom}_{\underline{C}}(C, I))$$

and this functor defines an equivalence of categories $\underline{C}^{o} \xrightarrow{\sim} TC(R)$. Furthermore, the conjugate category (§ 3) $\tilde{\underline{C}}$ of $\underline{C}$ is naturally equivalent to Dis(R).

Conversely, given any topologically coherent and topologically coperfect complete ring R, then $TC(R)^{o}$ is a locally noetherian category, and so in particular every projective of TC(R) is (uniquely) a direct product of indecomposable projectives. Further R is projective in TC(R), and every indecomposable projective is of the form Re, where e is a primitive idempotent, (i.e. e is not the sum of two non-zero, orthogonal idempotents). Finally, if R is topologically coherent and topologically coperfect (but not necessarily complete) then Dis(R) is the conjugate of a locally noetherian category, thus of the form $\mathrm{Dis}(\hat{R})$, where $\hat{R}$ is also complete (this sort of completion will be studied below).

PROOF: Everything has been proved except the assertion concerning the projective objects (R complete). But in $TC(R)^{o}$ every injective is an essentially unique sum of indecomposables, and so by duality every projective in TC(R) is a unique direct product of indecomposable ones. Further R is projective in TC(R) as is easily seen. Finally, let P be an indecomposable projective $\neq 0$ of TC(R), and let U be an open submodule of P such that P/U is coherent in Dis(R) and non-zero. Then the map $P \longrightarrow P/U$ is continuous, thus an epimorphism in TC(R) and so the projective envelope of P/U in TC(R) is a direct factor of P. Since P is indecomposable, $P \longrightarrow P/U$ is the projective envelope

[1] R is then complete, topologically coherent and topologically coperfect, cf. § 4.

of P/U. But P/U is coherent in Dis(R), thus in particular a finitely generated R-module, and so we have an epimorphism $\coprod_1^n R \to P/U \to 0$ in Mod(R), and this is a continuous map, thus an epimorphism in TC(R). It follows that P is a direct factor of $\coprod_1^n R$ (n finite) in TC(R), and since P is indecomposable, it is a direct factor of R (essential uniqueness of the decomposition of R), thus of the form Re, where e is a primitive idempotent. That e can be arbitrary (primitive) is easily seen.

COROLLARY 1.- Any linearly topologized topologically coherent and topologically coperfect complete ring R can be represented as the endomorphism ring with its natural topology of a big injective in a locally noetherian category $\underline{C}_R$, whose equivalence class is uniquely determined by R.

END OF THE PROOF OF THEOREM 4 (§ 4): Apply Corollary 1.

Remark.- Since a big injective in a locally noetherian category is not uniquely determined (a given indecomposable injective can occur an arbitrary number of times), we see that we can have $\underline{C}_R \xrightarrow{\sim} \underline{C}_{R'}$ without R being topologically isomorphic to R'. We will now study R more closely and also determine the degree of choice of R.

Recall that if $\underline{C}$ is an arbitrary Grothendieck category, $\underline{C} \xrightarrow{P} \mathrm{Spec}(\underline{C})$ the spectral category of $\underline{C}$ and I an arbitrary injective object of $\underline{C}$, then the kernel of the natural map

$$(17) \qquad \mathrm{Hom}_{\underline{C}}(I, I) \longrightarrow \mathrm{Hom}_{\mathrm{Spec}(\underline{C})}(P(I), P(I))$$

is exactly the Jacobson radical of the ring $\mathrm{Hom}_{\underline{C}}(I, I)$ [this radical will be denoted by $\mathrm{rad}_J(\mathrm{Hom}_{\underline{C}}(I, I))$] [59], [19], [51]. Furthermore, since (17) can be identified with the directed $\varinjlim$ of the epimorphisms (I is injective) $\mathrm{Hom}_{\underline{C}}(I, I) \longrightarrow \mathrm{Hom}_{\underline{C}}(V, I)$, $V \subset I$ essential subobject of I [51], (17) is an epimorphism. Thus the quotient ring $\mathrm{Hom}_{\underline{C}}(I, I)/\mathrm{rad}_J(\mathrm{Hom}_{\underline{C}}(I, I))$ is naturally isomorphic to

$\mathrm{Hom}_{\mathrm{Spec}(\underline{C})}(P(I), P(I))$ and thus it is a (von Neumann) regular right self-injective ring [51].

Now in particular, if $\underline{C}$ is locally noetherian then $\mathrm{Spec}(\underline{C})$ is discrete[1)] [27] i.e. of the form $\prod_{i \in S} \mathrm{Mod}(K_i)$ where S is the set of isomorphy classes of indecomposable injectives in $\underline{C}$, and where for $i \in S$, K_i is the skew field $\mathrm{Hom}_{\mathrm{Spec}(\underline{C})}(P(i), P(i))$. The functor P is easily made explicit on objects: to get $P(C)$, take the injective envelope $I(C)$ of C and decompose it into a direct sum of indecomposables. Then $P(C)$ is a set of vector spaces over K_i, $\{V_i\}_{i \in S}$ where the dimension over K_i of V_i is the number of times the indecomposable $i \in S$ occurs in $I(C)$. Thus in particular $I(C)$ is a big injective if and only if $\dim_{K_i} V_i > 0$, $i \in S$. Let us say (cf. [25]) that I is a sober injective if $\dim_{K_i} V_i = 1$, $i \in S$, i.e. if and only if every indecomposable occurs once and only once in $I(C)$. From what we just have said it follows that the fact that I is sober can be easily seen on the endomorphism ring of I, or more precisely on $\mathrm{Hom}_{\underline{C}}(I, I)/\mathrm{rad}_J(\mathrm{Hom}_{\underline{C}}(I, I))$ for this ring is naturally isomorphic to $\prod_{i \in S} \mathrm{End}_{K_i}(V_i)$.

Since by the Theorem 4 of § 4 every linearly topologized topologically coherent and topologically coperfect, complete ring is the endomorphism ring of a big injective in a locally noetherian category, we get:

COROLLARY 2.- *Let A be a linearly topologized, topologically coherent and topologically coperfect complete ring. Then $A/\mathrm{rad}_J(A)$ is a von Neumann regular right self-injective ring of the form $\prod_{i \in S} \mathrm{End}_{K_i}(V_i)$, where the skew fields K_i and the left K_i-vectorspaces V_i are uniquely determined.*

Remark 1.- $A/\mathrm{rad}_J(A)$ is also *left* self-injective if and only if

---

1) More generally $\mathrm{Spec}(\underline{C})$ is discrete if $\underline{C}$ has Krull dimension defined [25], and even more generally, if $\underline{C}$ is locally coirreducible [42] (this is even a necessary and sufficient condition for discreteness).

$\dim_{K_i} V_i < \infty, \forall i$ [52].

Remark 2.- Corollary 2 is probably also true if we leave out the coherence condition.

DEFINITION 4.- *With the notations of Corollary 2, we say that A is sober if* $\dim_{K_i} V_i = 1$, $i \in S$.

Using this definition we get the important

COROLLARY 3.- *The map* $A \longmapsto TC(A)^{o}$ *defines a one-one correspondence between the topological isomorphy classes of linearly topologized, topologically coherent, topologically coperfect and complete rings that are sober, and the equivalence classes of locally noetherian categories*[1].

If we say that two linearly topologized topologically coperfect, topologically coherent, complete rings A and B are "*topologically Morita equivalent*" if $TC(A) \xrightarrow{\sim} TC(B)$, then we obtain:

COROLLARY 4.- *Every linearly topologized, topologically coherent, topologically coperfect, complete ring* B *is topologically Morita equivalent to a uniquely determined such a ring* A *that is sober, and* B *can be explicitly described as a topological matrix ring over* A *as in* [38], [39].

The discrete rings of Corollary 4 are exactly the left coherent, right perfect rings (cf. Remark and example following Theorem 3 in § 3). For these rings, the topological Morita equivalence is the ordinary Morita equivalence [5], [25] as is easily seen. Thus:

COROLLARY 5.- *The map* $A \longmapsto TC(A)^{o}$ *defines a one-one correspondence between the Morita equivalence classes of left coherent, right perfect rings and the equivalence classes of locally noetherian categories that have a big noetherian injective* (i.e. every indecomposable injective is noetherian, and there are only a finite number of isomorphy classes of

---

1) This generalizes the result of Gabriel [24], [25], which says that the locally finite categories correspond to the pseudo-compact rings [25].

indecomposable injectives).

As we have remarked before in § 3, $\widetilde{\mathrm{Mod}\begin{pmatrix} \underset{\sim}{Q} & \underset{\sim}{C} \\ 0 & \underset{\sim}{Q} \end{pmatrix}} = \mathrm{TC}\begin{pmatrix} \underset{\sim}{Q} & \underset{\sim}{C} \\ 0 & \underset{\sim}{Q} \end{pmatrix}^{\circ}$ is an example of a locally noetherian category $\underline{C}$ satisfying the conditions of Corollary 5 that is not locally finite. If however $\underline{C}$ is also a module category, then $\underline{C}$ is locally finite [21].

Finally, let us observe that if $\underline{C}$ is locally noetherian, I a big injective of $\underline{C}$, $R = \mathrm{Hom}_{\underline{C}}(I, I)$, then the anti-equivalence

$$\underline{C} \ni C \longmapsto \mathrm{Hom}_{\underline{C}}(C, I) \in \mathrm{TC}(R)$$

transforms injective objects into projective objects in TC(R). <u>However</u>, these later objects are not necessarily projective and not even flat in Mod(R) in general. The situation can be completely clarified if we use the theory of Chase [13]:

THEOREM 7.- <u>With the notations and hypotheses of Theorem</u> 6, <u>the functor</u>

$$\underline{C} \ni C \longmapsto \mathrm{Hom}_{\underline{C}}(C, I) \in \mathrm{Mod}(R) \tag{18}$$

<u>transforms injective objects into projective</u> (resp. flat) <u>ones if and only if</u> R <u>is right coherent and left perfect</u> (resp. R is right coherent).

PROOF: Suppose first that (18) transforms injectives into projective (resp. flat) objects. Then since $R = \mathrm{Hom}_{\underline{C}}(I, I)$, we have that $\prod_K R = \mathrm{Hom}_{\underline{C}}(\coprod_K I, I)$ as a left R-module, and since $\coprod_K I$ is injective, we have by hypothesis that $\prod_K R$ is projective (resp. flat) as a left R-module for any K. Thus by Theorem 3.3. [resp. the left-right symmetric of Theorem 2.1] of [13] R is right coherent and left perfect [resp. right coherent (<u>cf.</u> also Bourbaki [9], p. 63 Exercise 12)].

Conversely, suppose that R is as in the last sentence, and let J be any injective in $\underline{C}$. Since I is a big injective, J is a direct factor of a suitable sum $\coprod_L I$, thus $\mathrm{Hom}_{\underline{C}}(J, I)$ is as a left R-module a direct factor of $\prod_L R$, thus projective (resp. flat) by the theorems of Chase (-Bourbaki) just cited, and so the Theorem 7 is completely proved.

PROBLEM: Is a left linearly topologized, left topologically coherent, left topologically coperfect, complete, left perfect, right coherent ring R discrete? (If we omit the condition left perfect, then R is <u>not</u> necessarily discrete as is easily seen.) If so, then the rings R of Theorem 7 would be discrete in the projective case. Such a discrete R is not necessarily artinian on any side: $\begin{pmatrix} \mathbb{Q} & \mathbb{C} \\ 0 & \mathbb{Q} \end{pmatrix}$ is coherent and perfect on both sides, but not artinian on any side.

We now pass to a second application of Theorem 5 above. Let A be linearly topologized, and $\tau = \mathrm{Art}(\mathrm{Dis}(A)) \subset \mathrm{Mod}(A)$ the full subcategory of Mod(A) formed by the discrete artinian modules. Then τ is evidently an artinian abelian category that is closed under formation of kernels and cokernels (it is even closed under the formation of subobjects and quotient objects in Mod(A)), and by Theorem 5 $\mathrm{Mod}_{\mathrm{Art}(\mathrm{Dis}(A))}(A)$ is the dual of a locally noetherian category in which dual the kernels, cokernels, etc. are the algebraic ones.[1)]If A with its linear topology, itself belongs to $\mathrm{Mod}_{\mathrm{Art}(\mathrm{Dis}(A))}(A)$ then A is exactly what is called a strictly linearly compact ring in [10], p. 111-112 [i.e. S.l.k. Ring in [38], [39] and a Leptin ring in [24]] and then $\mathrm{Mod}_{\mathrm{Art}(\mathrm{Dis}(A))}(A)$ can be identified with the category Lep(A) of Leptin modules over A (cf. Gabriel [24], [25]) which is thus the dual of a locally noetherian category [24], [25]. However in this case, it is <u>not</u> true in general that every projective indecomposable object in Lep(A) is of the form Ae, where e is a primitive idempotent (cf. examples in Gabriel [24]).

COROLLARY 6 (of theorem 6). <u>If</u> A <u>is a Leptin ring</u> (a strictly linearly compact ring) <u>then there exists a linearly topologized</u>, <u>topologically coherent</u>, <u>topologically coperfect and complete ring</u> B, <u>and an idempotent</u> e <u>in</u> B, <u>such that</u> A = eBe, <u>with its natural topology</u>. <u>In particular</u> A <u>is the endomorphism ring</u> (with its natural topology) <u>of an injective</u> (not necessarily big) <u>of a suitable locally noetherian category</u>.

---

1) The topology is the induced ... one.

Remark.- It is probably not true that every Leptin ring is topologically coherent (if this were true, then we could of course choose $B = A$, $e = 1$ in Corollary 6).

PROBLEM: If the ring R has a linear topology, so that $R \in \mathrm{Mod}_{\tau}(R)$ for a suitable $\tau \subset \mathrm{Dis}(R)$ (τ as in Theorem 5), then R is topologically coperfect and complete. Is the converse true? If not, characterize those rings that are obtainable in this manner.

§ 6. Topologically coherently completed tensor products.

This section will be used in § 7, notably for an explicit description of the Gabriel filtration of $TC(A)^{o}$, when A is a linearly topologized, topologically coherent, topologically coperfect and complete ring. We will introduce and study a notion of topologically coherently completed tensor product over A, denoted by $\widehat{\otimes}_A^c$. In the case when A is pseudo-compact [this is for instance the case if A is commutative - this follows from the theorem 3.4 of Chase [13]] then this tensor product coincides with the one introduced by Gabriel in [24] [1)] and denoted by $\widehat{\otimes}_A$ there, and it is well-known that this last tensor product generalizes the usual completed tensor product for modules over noetherian local rings, used for instance in [57] and in [10]. As a motivating (and - as we will see below in Theorem 8 and Remark 1 - an exhaustive) example for the introduction of $\widehat{\otimes}_A^c$, let $\underline{C}$ and $\underline{D}$ be two locally noetherian categories, A and B the associated topological rings, and $\mathrm{Lex}_{\sim \varinjlim}(\underline{C},\underline{D})$ the category of covariant left exact functors from $\underline{C}$ to $\underline{D}$ that commute with directed $\varinjlim$ [$\mathrm{Rex}_{\sim \varprojlim}(\ ,\ )$ is defined in an analogous way]. We have that (cf. § 2)

$$\mathrm{Lex}_{\sim \varinjlim}(\underline{C},\underline{D}) = \mathrm{Lex}(N(\underline{C}),\underline{D}) \quad [N(\underline{C}) = \text{the category of noetherian objects of } \underline{C}.]$$

and that

---

1) Cf. also [26] for the commutative case.

$$\mathrm{Lex}_{\mathcal{N}\ \varinjlim}(\underline{C},\underline{D})^{o} \xrightarrow{\sim} \mathrm{Rex}_{\mathcal{N}\ \varprojlim}(TC(A),\ TC(B)).$$

If T is a functor belonging to this last category, then not only does T(A) belong to TC(B), but this T(A) also has a natural <u>right</u> A-module structure, compatible with the <u>left</u> B-module structure: right multiplication by $\lambda \in A$ is a continuous map $A \xrightarrow{.\lambda} A$ that is transformed by T into a map $T(A) \xrightarrow{T(.\lambda)} T(A)$ in TC(B). We will write this operation $T(.\lambda)$ as $.\lambda$, and we will now prove that this right operation of A on M = T(A) is continuous (so that M is in fact a topological B-A-bimodule). Using the formula

$$m\cdot\lambda - m_o\cdot\lambda_o = (m - m_o)\cdot(\lambda - \lambda_o) + m_o\cdot(\lambda - \lambda_o) + (m - m_o)\cdot\lambda_o$$

we see that it is sufficient to prove the following three results:

1) The natural map $M \times A \longrightarrow M$ is continuous at (0,0).
2) For each $m_o \in M$, the map $A \longrightarrow M$ defined by $\lambda \longmapsto m_o\cdot\lambda$ is continuous at 0.
3) For each $\lambda_o \in A$, the map $M \longrightarrow M$ defined by $m \longmapsto m\cdot\lambda_o$ is continuous at 0.

We have just seen that 3) is true. To prove 1) and 2), let $\{\mathfrak{A}_\alpha\}$ be the directed decreasing set of open left ideals of A, that belong to TC(A), and let U be an open B-submodule of M = T(A) such that M/U is in τ. Then M/U is an artinian object of TC(B), and thus the directed decreasing family of subobjects of M/U in TC(B):

$$T_\alpha = \mathrm{Im}(T(\mathfrak{A}_\alpha) \longrightarrow T(A) \longrightarrow T(A)/U)$$

has a minimum element $T_{\alpha^*} = T_\alpha$, $\alpha \geq \alpha^*$. But the directed $\varprojlim$ are exact in TC(B) and $\varprojlim T(\mathfrak{A}_\alpha) = T(\varprojlim \mathfrak{A}_\alpha) = 0$, and so

$$\mathrm{Im}(T(\mathfrak{A}_{\alpha^*}) \longrightarrow T(A) \longrightarrow T(A)/U) = 0$$

which implies that $M.\mathfrak{A}_{\alpha^*} \subset U$, and this proves both 1) and 2) and even the (apparently [1)]) stronger result that the map $M \times A \longrightarrow M$ is left equicontinuous (cf. § 8). More precisely:

[1)] That the equicontinuity property is only apparently stronger <u>here</u> can be proved in much the same way as the lemma 0.3.1 on p. 71 of [26] is proved (it is essential that $M \in TC(B)$ as a left B-module, cf. § 8).

THEOREM 8. - Let A and B be as above. Then the functor

$$\mathrm{Rex}_{\underleftarrow{\mathcal{N}\,\lim}}(TC(A),\ TC(B)) \ni T \longmapsto T(A) \in {}_B TCT_A$$

[where ${}_B TCT_A$ is the category whose objects are those topological B-A-bimodules M, that belong to TC(B) as left B-modules, and whose morphisms are the continuous bimodule maps] is an equivalence of categories. Given $M \in {}_B TCT_A$, the corresponding functor $TC(A) \longrightarrow TC(B)$ will be denoted by $M \hat{\otimes}_A^c -$, for the following reason: For $N \in TC(A)$, we have a natural equivalence of functors:

$$\mathrm{Hom}_{TC(B)}(M \hat{\otimes}_A^c N,\ V) \xrightarrow{\sim} \mathrm{Biltop}_B(M,\ N;\ V), \quad V \in \mathrm{Coh}(\mathrm{Dis}(B)),$$

where $\mathrm{Biltop}_B(M, N; V)$ is the set of continuous maps $M \times N \to V$ that define B-linear maps $M \otimes_A N \longrightarrow V$.

INDICATIONS OF THE PROOF OF THEOREM 8: We have seen in § 5 that

$$TC(B) \ni L \longmapsto \mathrm{Hom}_{TC(B)}(L,\ -) \in \mathrm{Lex}(\mathrm{Coh}(\mathrm{Dis}(B)),\ \underline{Ab})$$

is anti-equivalence of categories. Now let M be in ${}_B TCT_A$. Then $\mathrm{Biltop}_B(M, N; -) \in \mathrm{Lex}(\mathrm{Coh}(\mathrm{Dis}(B)), \underline{Ab})$ and so it is of the form $\mathrm{Hom}_{TC(B)}(L, -)$ for a unique L $[- \in \mathrm{Coh}(\mathrm{Dis}(B))]$. This L defines $M \hat{\otimes}_A^c N$, and one verifies that the Theorem 8 is true.

Remark 1.- Of course $\mathrm{Biltop}_B(M, N; -) \in \mathrm{Lex}(\mathrm{Coh}(\mathrm{Dis}(B)), \underline{Ab})$ even if M satisfies weaker conditions, and so we have a $M \hat{\otimes}_A^c N$ in this case too. The Theorem 8 implies that there exists a "completion" $\hat{M} \in {}_B TCT_A$ of M such that $M \hat{\otimes}_A^c N \xrightarrow{\sim} \hat{M} \hat{\otimes}_A^c N$ (N can also satisfy weaker conditions ...). One verifies that $\hat{\otimes}_A^c = \hat{\otimes}_A$ in the pseudo-compact case [24], [26].

Remark 2.- If $\underline{D} = \mathrm{Mod}(\underset{\sim}{Z})$, $B = \mathrm{End}_{\underset{\sim}{Z}}(\underset{\sim}{Q} \amalg \underset{\sim}{Q}/\underset{\sim}{Z})$, then

$$\mathrm{Rex}_{\underleftarrow{\mathcal{N}\,\lim}}(TC(A),\ TC(B)) = \mathrm{Dis}(A)^{\circ}$$

and if $\underline{C}$ is also locally finite, then $\mathrm{Dis}(A)^{\circ} \xrightarrow{\sim} TC(A^{*})$ $[\xrightarrow{\sim} PC(A^{*})]$, where A^{*} is a certain pseudo-compact ring introduced in [24],[25], and called the dual of A. Thus we have ${}_B TCT_A \xrightarrow{\sim} TC(A^{*})$, and this gives some indication that A^{*} should be obtainable as a suitable topological tensor product of B and of A^{opp}. We have not tried to pursue this further, nor have we tried to see whether there is a suitable linearly topologized ring A^* in the general locally noetherian case (then A^* is of course not necessarily topologically coperfect and topologically coherent).

§ 7.- <u>Explicit study of the dual and the conjugate of the Gabriel filtration of a locally noetherian category</u>.

Let $\underline{C}$ be a locally noetherian category. We will now use the representation $\underline{C} \xrightarrow{\sim} TC(A)^{o}$ of Theorem 6 to describe explicitly the Gabriel filtration of $\underline{C}$, and how $\underline{C}$ is built up by "extensions" of locally finite categories. We will have the most complete results in the case when $\underline{C}$ is stable, i.e. when the Gabriel filtration is stable under injective envelopes (cf. § 8 och § 9), but here we start first with some results in the general case.

If $\underline{E}$ is an AB 5^*-category with a family of cogenerators [30] (i.e. the dual of a Grothendieck category), then a subcategory $\underline{F}$ of $\underline{E}$ will be called a <u>coclosed</u> category if it is the dual of a closed subcategory [25] of $\underline{E}^{o}$. It is equivalent to say that $\underline{F}$ is a full subcategory of $\underline{E}$ that is stable under the formation of subobjects, quotient objects and $\varprojlim$. In the same way we introduce the notion of a colocalizing subcategory, the notion of product of two coclosed subcategories of $\underline{E}$ (cf. [25], p. 395) etc. If $\underline{F}$ is a coclosed subcategory of $\underline{E}$, then the inclusion functor $\underline{F} \xrightarrow{i_*} \underline{E}$ has a <u>left</u> adjoint i^* and if $\underline{F}$ is a colocalizing subcategory, then the quotient category $\underline{E}/\underline{F}$ can be formed and the natural functor $\underline{E} \xrightarrow{j^*} \underline{E}/\underline{F}$ has a <u>left</u> adjoint $j_!$ that is a full embedding, and so we have an "exact sequence"

$$0 \longrightarrow \underline{F} \underset{i_*}{\overset{i^*}{\rightleftarrows}} \underline{E} \underset{j^*}{\overset{j_!}{\rightleftarrows}} \underline{E}/\underline{F} \longrightarrow 0$$

Now we will see that in case $\underline{E} = TC(A)$ [i.e. E is the dual of a locally noetherian category], then all these categories and functors can be described in a very explicit manner. Our results generalize and complete those of Gabriel [25], p. 400.

THEOREM 9. - Let TC(A) be as before. The map that to each bilateral ideal $\mathfrak{a}$ of A that belongs to TC(A) as a left A-module, associates the full subcategory $\underline{F} = TC(A/\mathfrak{a}) \xrightarrow{i_{*}} TC(A)$ (i_{*} is the natural embedding), defines a one-one correspondence between these ideals $\mathfrak{a}$ and the coclosed subcategories $\underline{F}$ of TC(A). With the notations above and of § 6 we have $i_{*}i^{*}(M) = A/\mathfrak{a} \hat{\otimes}_A^c M$. If $\underline{F}_1$ and $\underline{F}_2$ are two coclosed subcategories corresponding to $\mathfrak{a}_1$ and $\mathfrak{a}_2$, and if $\underline{F}_1 \cdot \underline{F}_2$ is the product subcategory [25], then this coclosed subcategory corresponds to the bilateral ideal $\overline{\mathfrak{a}_1 \cdot \mathfrak{a}_2}^c$, i.e. the smallest left ideal of A that is in TC(A) and that contains $\mathfrak{a}_1 \cdot \mathfrak{a}_2$ (this smallest ideal is bilateral and it is also the image of the multiplication map $\mathfrak{a}_1 \hat{\otimes}_A^c \mathfrak{a}_2 \longrightarrow A$). In particular $\underline{F}$ is a colocalizing subcategory if and only if $\overline{\mathfrak{a}^2}^c = \mathfrak{a}$, and in this case the ingredients of the exact sequence

$$0 \longrightarrow TC(A/\mathfrak{a}) \underset{i_{*}}{\overset{i^{*}}{\rightleftarrows}} TC(A) \underset{j^{*}}{\overset{j_{!}}{\rightleftarrows}} TC(A)/TC(A/\mathfrak{a}) \longrightarrow 0$$

can be made explicit as follows:

1) $j_{!}j^{*}(M) = \mathfrak{a} \hat{\otimes}_A^c \mathfrak{a} \hat{\otimes}_A^c M$;

2) $TC(A)/TC(A/\mathfrak{a})$ can be identified with the full subcategory of TC(A), formed by those $M \in TC(A)$ such that the multiplication map $\mathfrak{a} \hat{\otimes}_A^c M \longrightarrow M$ is an isomorphism ($j_{!}$ is then the inclusion functor);

3) The natural exact sequence

$$j_{!}j^{*}(M) \longrightarrow M \longrightarrow i_{*}i^{*}(M) \longrightarrow 0$$

can be identified with the natural tensor product exact sequence

$$\mathfrak{a} \hat{\otimes}_A^c \mathfrak{a} \hat{\otimes}_A^c M \longrightarrow M \longrightarrow A/\mathfrak{a} \hat{\otimes}_A^c M \longrightarrow 0$$

$\mathfrak{a}\hat{\otimes}^c_A \mathfrak{a}\hat{\otimes}^c_A M \longrightarrow \mathfrak{a}\hat{\otimes}^c_A M$ is onto).

Finally, if $\overline{A/\mathfrak{a}}$ is the projective envelope of $A/\mathfrak{a}$ in TC(A), then $A = \overline{A/\mathfrak{a}} \amalg Ae$, where e is an idempotent of A, and if $\overline{\mathfrak{a}^2}^c = \mathfrak{a}$ then eAe is with its natural topology a topologically coherent (topologically coperfect is clear) complete ring, and (using 2)) $TC(A)/TC(A/\mathfrak{a})$ is naturally equivalent to TC(eAe), by means of the functor $M \longmapsto Hom_{TC(A)}(Ae, M)$. Also $TC(A/\mathfrak{a}) \hookrightarrow TC(A)$ is stable under projective envelopes if and only if $A/\mathfrak{a}$ is projective in TC(A) (gen. case). [In this last case, we can choose e such that $\mathfrak{a} = A(1-e)$, and then $(1-e)Ae = 0$.]

The proof of this theorem is entirely based on the theory of § 6, and the general functorial properties of $j_!$, j^*. (Compare also [48].)

Now that we have Theorem 9 and § 5 - § 6, we can easily translate the definitions and results of [25], p. 382 ... into our dual language:

COROLLARY 1.- Let A be as before. Then among the left ideals $\mathfrak{a}$ of A, $\mathfrak{a} \in TC(A)$ such that $A/\mathfrak{a}$ is a pseudo-compact left A-module, there is a smallest one $\mathfrak{a}_o$, and this $\mathfrak{a}_o$ is bilateral and $\overline{\mathfrak{a}_o^2}^c = \mathfrak{a}_o$, so that $A/\mathfrak{a}_o$ is a left pseudo-compact ring, and $TC(A/\mathfrak{a}_o) \hookrightarrow TC(A)$ is a colocalizing subcategory. There is also a smallest left ideal of A, $\mathfrak{a}_1 \in TC(A)$ such that the image of $A/\mathfrak{a}_1$ in $TC(A)/TC(A/\mathfrak{a}_o)$ is pseudo-compact (this can be expressed with the projective envelope of $A/\mathfrak{a}_o$ in TC(A) too.) This ideal $\mathfrak{a}_1$ is also bilateral and $\overline{\mathfrak{a}_1^2}^c = \mathfrak{a}_1$, and $\mathfrak{a}_1 \subset \mathfrak{a}_o$. Continuing in this manner, we obtain a decreasing filtration of A by bilateral, left TC ideals $\{\mathfrak{a}_\alpha\}$ of A such that $\overline{\mathfrak{a}_\alpha^2}^c = \mathfrak{a}_\alpha$:

$$A \supset \mathfrak{a}_o \supset \mathfrak{a}_1 \supset \dots \supset \mathfrak{a}_\alpha \supset \dots$$

[If α has no predecessor, i.e. is a limit ordinal, then we put $\mathfrak{a}_\alpha$ = the biggest left ideal in TC(A) such that $\mathfrak{a}_\alpha \subset \mathfrak{a}_\beta$, $\beta < \alpha$ [1].] Then there is a unique smallest ordinal α^* such that $\mathfrak{a}_{\alpha^*} = 0$ and $\mathfrak{a}_\alpha \neq 0$, $\alpha < \alpha^*$. This α^* coincides with the Krull dimension of $TC(A)^o$, and

$$TC(A/\mathfrak{a}_o)^o \subset TC(A/\mathfrak{a}_1)^o \subset \ldots \subset TC(A/\mathfrak{a}_\alpha)^o \subset \ldots \subset TC(A)^o$$

coincides with the Gabriel filtration of $TC(A)^o$.

COROLLARY 2.- The Gabriel filtration of $TC(A)^o$ is stable under injective envelopes if and only if each $\mathfrak{a}_\alpha = Ae_\alpha$, where e_α is an idempotent of A [then $e_\alpha A(1-e_\alpha) = 0$].

Before we go over to a more detailed study of the stable case in the next two sections, we will first see how the conjugate categories fit into our picture.

THEOREM 10.- Let $\underline{C}$ be a locally noetherian category, A its associated topological ring and $\underline{\tilde{C}}$ = Dis(A) its conjugate. Then the map $\underline{F} \mapsto \underline{\tilde{F}}$ defines a one-one correspondence between the closed subcategories of $\underline{C}$ and closed subcategories of $\underline{\tilde{C}}$ [and $\underline{F} \mapsto \underline{\tilde{F}}$ can be identified with $\mathfrak{a} \mapsto$ Dis$(A/\mathfrak{a})$]. The localizing subcategories correspond and $\widetilde{\underline{C}/\underline{F}}$ = $= \underline{\tilde{C}}/\underline{\tilde{F}}$. In particular the Gabriel filtration $\underline{C}_o \subset \underline{C}_1 \subset \ldots \subset \underline{C}_\alpha \subset \ldots \subset \underline{C}$ gives rise to a filtration of $\underline{\tilde{C}}$ = Dis(A):

$$(19) \qquad \underline{\tilde{C}}_o \subset \underline{\tilde{C}}_1 \subset \ldots \subset \underline{\tilde{C}}_\alpha \subset \ldots \subset \underline{\tilde{C}}\ ,$$

which can be identified with

$$\mathrm{Dis}(A/\mathfrak{a}_o) \subset \mathrm{Dis}(A/\mathfrak{a}_1) \subset \ldots \subset \mathrm{Dis}(A/\mathfrak{a}_\alpha) \subset \ldots \subset \mathrm{Dis}(A).$$

Furthermore $\underline{\tilde{C}}_o$ is the smallest localizing subcategory of $\underline{\tilde{C}}$ that

---

1) By § 5 we have that $\mathfrak{a}_\alpha = \bigcap_{\beta < \alpha} \mathfrak{a}_\beta$

contains the simple and coherent objects [1), $\widetilde{\underline{C}}_1$ is the smallest localizing subcategory of $\widetilde{\underline{C}}$ that contains those objects whose images in $\widetilde{\underline{C}}/\widetilde{\underline{C}}_o$ are coherent and simple etc. From the results above follows that $\widetilde{\underline{C}}_{i+1}/\widetilde{\underline{C}}_i$ is locally finite, thus of the form $PC(A_i^{*})^o$ (cf. § 6, Remark 3).

PROBLEM 1: Can an A^{*} be built up from these A_i^{*} (at least in the stable case)? (Cf. § 6, Remark 3.)

PROBLEM 2: Does every Grothendieck category $\underline{D}$ of Krull dimension zero admit a filtration similar to (19) above?

§ 8. Generalized triangular matrix rings with a linear topology, and classification of stable extensions of locally noetherian categories.

Let $\underline{C}$ be a Grothendieck category, $\underline{D}$ a localizing subcategory, $\underline{C}/\underline{D}$ the quotient category and

$$(20) \qquad 0 \longrightarrow \underline{D} \xrightarrow{i_*} \underline{C} \xrightarrow{j^*} \underline{C}/\underline{D} \longrightarrow 0$$

the sequence of natural functors. Recall that here i_* and j^* have right adjoints $i^!$ and j_* that are full and faithful, and $j^*(C) = 0 \iff C = i_*(D)$. We will say that (20) is an extension of $\underline{C}/\underline{D}$ by $\underline{D}$. Recall that Gabriel has proved that any locally noetherian category can be obtained in a canonical way by means of successive extensions of locally finite ones [for more details see [25], Chap. IV (cf. also § 7)] and that he has explicit results about the structure of the locally finite ones (some of which we have extended here to the locally noetherian case). However, as far as we know, the problem to classify the extensions (20) of locally finite categories or more generally of locally noetherian categories, has not been dealt with in the literature, and we will here give the rather complete results that we have obtained

---

1) The smallest localizing subcategory of $\widetilde{\underline{C}}$ that contains all the simple objects of $\widetilde{\underline{C}}$, is of course $\widetilde{\underline{C}}$ itself, since $\widetilde{\underline{C}}$ is of Krull dimension zero (§ 3).

in the stable case. [We say that (20) is a stable extension, or that $\underline{D}$ is a stable subcategory of $\underline{C}$ if the injective envelope in $\underline{C}$ of every object in $\underline{D}$, is still in $\underline{D}$.] The stable case occurs frequently "in practice": It is for example well-known [25] that if X is a noetherian prescheme, $\underline{C}$ = Qcoh(X), the category of quasicoherent sheaves over X, then every localizing subcategory of $\underline{C}$ is stable.

Thus consider now an "exact sequence"

$$(21)\qquad 0 \longrightarrow \underline{D} \xrightarrow{i_*} \underline{C} \xrightarrow{j^*} \underline{E} \longrightarrow 0$$

($\underline{D}$ a localizing subcategory of $\underline{C}$, and $\underline{E}$ the quotient category), where $\underline{C}$ is locally noetherian (then $\underline{D}$ and $\underline{E}$ are so too), and let I be a big injective of $\underline{C}$. Suppose that the sequence is stable. Then we have a canonical decomposition of I ([25], p. 375 cor. 2-3)

$$(22)\qquad I = i_*(I_0) \amalg j_*(I_1)$$

where I_0 is a big injective in $\underline{D}$ and I_1 is a big injective in $\underline{E}$. Since $j^* i_* = 0$, we get $\mathrm{Hom}_{\underline{C}}(i_*(I_0), j_*(I_1)) = 0$, and so from (22) we obtain a direct decomposition of abelian groups

$$(23)\qquad A = \mathrm{Hom}_{\underline{C}}(I, I) = \mathrm{Hom}_{\underline{D}}(I_0,I_0) \amalg \mathrm{Hom}_{\underline{C}}(j_*(I_1),i_*(I_0)) \amalg \mathrm{Hom}_{\underline{E}}(I_1,I_1)$$

(i_* and j_* are fully faithful).

However, it is possible to rewrite (23) so that the ring structure of A becomes apparent. If we put $A_0 = \mathrm{Hom}_{\underline{D}}(I_0,I_0)$, $A_1 = \mathrm{Hom}_{\underline{E}}(I_1,I_1)$ and $M = \mathrm{Hom}_{\underline{C}}(j_*(I_1), i_*(I_0))$, then A_0 and A_1 are rings and M is in a natural way a left A_0-module and a right A_1-module and these structures are compatible (we will denote this by $M = {}_{A_0}M_{A_1}$).

Now, if A_0, A_1, ${}_{A_0}M_{A_1}$ are arbitrary, we will denote by $\begin{pmatrix} A_0 & {}_{A_0}M_{A_1} \\ 0 & A_1 \end{pmatrix}$ the ring whose elements are the triples (a_0,m,a_1) [more suggestively written as $\begin{pmatrix} a_0 & m \\ 0 & a_1 \end{pmatrix}$] $(a_0 \in A_0,\ m \in M,\ a_1 \in A_1)$, where addition is defined by componentwise addition, and where multiplication is defined by "matrix multiplication"

$$\begin{pmatrix} a_o & m \\ 0 & a_1 \end{pmatrix}\begin{pmatrix} a_o' & m' \\ 0 & a_1' \end{pmatrix} = \begin{pmatrix} a_o a_o' & a_o m' + m a_1' \\ 0 & a_1 a_1' \end{pmatrix}$$ (This has a sense, since $M = {}_{A_o}M_{A_1}$.)

This kind of generalized triangular matrix ring was first introduced by S.U. CHASE in [14] (cf. also [33]), and it is now easy to see that the assertion (23) can be made more precise by saying that we have a natural ring isomorphism

(24) $$A = \mathrm{Hom}_{\underline{C}}(I, I) \xrightarrow{\sim} \begin{pmatrix} \mathrm{Hom}_{\underline{D}}(I_o, I_o) & M \\ 0 & \mathrm{Hom}_{\underline{E}}(I_1, I_1) \end{pmatrix}$$

where $M = \mathrm{Hom}_{\underline{C}}(j_*(I_1), i_*(I_o))$.

Now in (24), to give the rings $A_o = \mathrm{Hom}_{\underline{D}}(I_o, I_o)$ and $A_1 = \mathrm{Hom}_{\underline{E}}(I_1, I_1)$ (with their natural topology), is the same as to give $\underline{D}$ and $\underline{E}$, and thus all information about the extension (21) should be contained in the A_o-A_1-bimodule M (this module also has a linear topology ...). Thus the problem arises to characterize those bimodules M that arise from stable extensions. This problem will be solved completely in what follows, but first we will have to develop several preliminaries (some of them perhaps of independent interest) about generalized triangular matrix rings.

Let $A = \begin{pmatrix} A_o & {}_{A_o}M_{A_1} \\ 0 & A_1 \end{pmatrix}$ be an arbitrary generalized triangular matrix ring. We wish first to determine all the left ideals of A. For that we start with some examples of such ideals. It is clear that

$\begin{pmatrix} A_o & {}_{A_o}M_{A_1} \\ 0 & 0 \end{pmatrix}$ where A operates to the left through matrix multiplication

is a left ideal of A, and this ideal can, in fact, be described as the left A_o-module $A_o \amalg M$ $(M = {}_{A_o}M_{A_1})$ on which A operates through A_o. It follows that every left A_o-submodule $V \subset A_o \amalg M$ defines a left ideal of

A, that we will denote by $\begin{pmatrix} & V \\ 0 & 0 \end{pmatrix}$ [note that with this notation we can write $\begin{pmatrix} A_0 & M \\ 0 & 0 \end{pmatrix} = \begin{pmatrix} & A_0 \amalg M \\ 0 & 0 \end{pmatrix}$]. On the other hand, if $\mathfrak{a}_1$ is a left ideal of A_1, then the left ideal of A generated by $\begin{pmatrix} 0 & 0 \\ 0 & \mathfrak{a}_1 \end{pmatrix}$ is exactly $\begin{pmatrix} 0 & M.\mathfrak{a}_1 \\ 0 & \mathfrak{a}_1 \end{pmatrix}$. Finally any sum of these two types of left ideals of A: $\begin{pmatrix} & V \\ 0 & 0 \end{pmatrix} + \begin{pmatrix} 0 & M.\mathfrak{a}_1 \\ 0 & \mathfrak{a}_1 \end{pmatrix}$ is evidently a left ideal of A, which we will write in the form [1] $\begin{pmatrix} & W \\ 0 & \mathfrak{a}_1 \end{pmatrix}$, where $W = V + M.\mathfrak{a}_1 \subset A_0 \amalg M$ is an A_0-submodule of $A_0 \amalg M$ such that $M.\mathfrak{a}_1 \subset W$. Except for this last restriction, $\mathfrak{a}_1 \subset A_1$ and $W \subset A_0 \amalg M$ can be completely arbitrary:

PROPOSITION 13.- Let $A = \begin{pmatrix} A_0 & M \\ 0 & A_1 \end{pmatrix}$ be an arbitrary generalized triangular matrix ring. The left ideals of A are exactly the subsets of A of the form $\begin{pmatrix} & V \\ 0 & \mathfrak{a}_1 \end{pmatrix}$, where V is an A_0-submodule of $A_0 \amalg M$, and where $\mathfrak{a}_1$ is a left ideal of A_1 such that $M.\mathfrak{a}_1 \subset V$, and where A operates on $\begin{pmatrix} & V \\ 0 & \mathfrak{a}_1 \end{pmatrix}$ by left matrix multiplication.

PROOF: Compute directly the left ideal of A, generated by a set of generators.

Remark.- If $\mathfrak{a}$ is a left ideal of A, then the corresponding $V_{\mathfrak{a}} \subset A_0 \amalg M$ and $\mathfrak{a}_1 \subset A_1$ are uniquely determined by $\mathfrak{a}$. Furthermore $A/\mathfrak{a}$ can be identified with [1] $\begin{pmatrix} & (A_0 \amalg M)/V_{\mathfrak{a}} \\ 0 & A_1/\mathfrak{a}_1 \end{pmatrix}$, where A operates by left matrix multiplication which is well-defined since $M.\mathfrak{a}_1 \subset V_{\mathfrak{a}}$. This gives rise to an exact sequence of left A-modules

$$(25) \qquad 0 \longrightarrow (A_0 \amalg M)/V_{\mathfrak{a}} \longrightarrow A/\mathfrak{a} \longrightarrow A_1/\mathfrak{a}_1 \longrightarrow 0$$

where A operates through A_0 on $(A_0 \amalg M)/V_{\mathfrak{a}}$, and through A_1 on $A_1/\mathfrak{a}_1$. This sequence, which will be very useful below, does not split in general, not even in the case $V_{\mathfrak{a}} = 0$, $\mathfrak{a}_1 = 0$ (i.e. $\mathfrak{a} = 0$)!

---

[1] We use here an extension of the notation introduced above.

COROLLARY.- *With the notations above,* $\mathfrak{a} \subset A$ *is a finitely generated left ideal, if and only if* $\mathfrak{a}_1$ *and* $V_{\mathfrak{a}}/M.\mathfrak{a}_1$ *are finitely generated as left* A_1*- and left* A_0*-modules respectively.*

We now turn to the study of linear topologies on generalized triangular matrix rings [1]. Combining Proposition 3 of § 4 with Proposition 13, we obtain:

PROPOSITION 14.- *To give a (left) linear topology on the ring* $A = \begin{pmatrix} A_0 & M \\ 0 & A_1 \end{pmatrix}$ *is the same as to give a (left) linear topology on the ring* A_1 *and a linear topology on the left* A_0*-module* $A_0 \amalg M$ *such that the maps:*

(i) $(A_0 \amalg M) \times (A_0 \amalg M) \longrightarrow A_0 \amalg M, \ \{(a_0,m),(a_0',m')\} \longmapsto (a_0a_0', a_0m')$

and

(ii) $(A_0 \amalg M) \times A_1 \longrightarrow A_0 \amalg M, \ \{(a_0,m), a_1\} \longmapsto (0, ma_1)$

are continuous, and such that furthermore the map (ii) *is left equicontinuous in the sense that to each open* A_0*-submodule* V *of* $A_0 \amalg M$ *there is an open left ideal* $\mathfrak{a}_1$ *of* A_1 *such that* $(A_0 \amalg M).\mathfrak{a}_1 \subset V$. *The topology on* A *is then the product of the topologies on* A_1 *and on* $A_0 \amalg M$.

Those triangular matrix rings that are of interest for us in connection with locally noetherian categories will have split linear topologies in the sense of the following definition.

DEFINITION 4.- *We say that a left ideal* $\mathfrak{a}$ *of* $A = \begin{pmatrix} A_0 & M \\ 0 & A_1 \end{pmatrix}$ *is a split ideal if* $\mathfrak{a}$ *is of the form* $\begin{pmatrix} \mathfrak{a}_0 & U \\ 0 & \mathfrak{a}_1 \end{pmatrix}$, *where* $\mathfrak{a}_i \subset A_i$ $(i = 0,1)$ *are left ideals,* $U \subset M$ *is a left* A_0*-submodule and* $M.\mathfrak{a}_1 \subset U$. *A left linear topology on the ring* A *is said to be a split linear topology if* A *has a fundamental system of open split left ideals.*

---

[1] Recall that in this paper, all topologies studied on rings are supposed to be compatible with the ring structure.

Remark 1.- To say that the topology on the linearly topologized ring $A = \begin{pmatrix} A_o & M \\ 0 & A_1 \end{pmatrix}$ is a split topology is the same as to say that the left A_o-linear topology on $A_o \amalg M$ (Proposition 14) is the product topology of such topologies on A_o and M. Thus Proposition 14 implies that to give a left linear split topology on the ring A is the same as to give left linear topologies on the rings A_o and A_1 and a linear topology on the left A_o-module M such that M becomes a topological A_o-A_1-bimodule such that the operation $M \times A_1 \longrightarrow M$ is left equicontinuous. The topology on A is then the product of the topologies on A_o, A_1 and M, and so A is in particular complete if and only if A_o, A_1 and M are so.

Remark 2.- The discrete topology on A is a split linear topology. Less trivial examples are given by:

THEOREM 11.- *Let $A = \begin{pmatrix} A_o & M \\ 0 & A_1 \end{pmatrix}$ be a left linearly topologized ring that is topologically coherent, topologically coperfect and complete. Then the topology on A is a split linear topology, and we even have a fundamental system $\{\mathfrak{A}\}$ of split open left ideal neighbourhoods of 0 in A such that every $A/\mathfrak{A}$ is coherent in* Dis(A).

PROOF: We know by § 5 that $\underline{C} = TC(A)^o$ is a locally noetherian category, that A corresponds in a natural way to a big injective I in $\underline{C}$, that the topology on $A = Hom_{\underline{C}}(I, I)$ is defined by the left ideals $l(N) = Hom_{\underline{C}}(I/N, I)$, where $N \subset I$ is a noetherian subobject, and that every $A/l(N)$ is coherent in Dis(A). Put $e = \begin{pmatrix} 1 & 0 \\ 0 & 0 \end{pmatrix}$. Then $Ae = \begin{pmatrix} A_o & 0 \\ 0 & 0 \end{pmatrix}$, $A(1-e) = \begin{pmatrix} 0 & M \\ 0 & A_1 \end{pmatrix}$ and the decomposition $A = Ae \amalg A(1-e)$ corresponds to a decomposition $I = I_o \amalg I_1$ of I in $\underline{C}$. Furthermore

$$(26) \qquad Hom_{\underline{C}}(I_o, I_1) = Hom_{TC(A)}(A(1-e), Ae).$$

But if f is an arbitrary left A-linear map $A(1-e) \longrightarrow Ae$, then $f(1-e) = \lambda e$, thus $f(1-e) = (1-e)f(1-e) = (1-e)\lambda e = 0$, since $(1-e)Ae = 0$ by direct matrix computation. Thus $f = 0$ and *a fortiori* the right member of (26) is 0, so that $\mathrm{Hom}_{\underline{C}}(I_o, I_1) = 0$.

Now let $\underline{C}_o$ be the localizing subcategory of $\underline{C}$, formed by those C such that $\mathrm{Hom}_{\underline{C}}(C, I_1) = 0$. I claim that $\underline{C}_o$ is stable under the formation of injective envelopes in $\underline{C}$, and that I_o is a big injective in $\underline{C}_o$. But if $C \in \underline{C}_o$ were such that its injective envelope (in $\underline{C}$) $I(C)$ did not belong to $\underline{C}_o$, then $I(C)$ would have an indecomposable direct factor $I_\alpha \neq 0$ that occurs as a direct factor of I_1 ($I_o \amalg I_1$ is a big injective of $\underline{C}$ and $I_o \in \underline{C}_o$). But the pullback diagram

$$\begin{array}{ccc} C & \hookrightarrow & I(C) \\ \uparrow & & \uparrow \\ C_\alpha & \hookrightarrow & I_\alpha \end{array}$$

gives rise to an object $C_\alpha \neq 0$ that belongs to $\underline{C}_o$. We have a monomorphism $I_\alpha \xrightarrow{i_\alpha} I_1$, and since $C_\alpha \in \underline{C}_o$ the composition map $C_\alpha \hookrightarrow I_\alpha \xhookrightarrow{i_\alpha} I_1$ is zero and this gives a contradiction.

Thus we have a stable exact sequence

$$0 \longrightarrow \underline{C}_o \longrightarrow \underline{C} \underset{j_*}{\overset{j^*}{\rightleftarrows}} \underline{C}/\underline{C}_o \longrightarrow 0$$

where I_o is a big injective of $\underline{C}_o$, $I = I_o \amalg I_1$ and $I_1 \xrightarrow{\sim} j_* j^*(I_1)$. Now if N is a noetherian subobject of I, then since both I_o and I_1 are directed unions of noetherian subobjects, we get that there are noetherian subobjects $N_o \subset I_o$ and $N_1 \subset I_1$ such that $N \subset N_o \amalg N_1$, and thus $l(N) \supset l(N_o \amalg N_1)$. It is clear that this last ideal is a split open left ideal of A, and that $A/l(N_o \amalg N_1)$ is coherent in $\mathrm{Dis}(A)$ and so the Theorem 11 is completely proved, as well as the

COROLLARY.- <u>With the notations and hypotheses of Theorem</u> 11, A_o <u>and</u> A_1 <u>are topologically coherent, topologically coperfect and complete for their natural linear topologies</u>.

<u>Remark</u>.- We will later determine exactly those conditions on M that assure that A is topologically coherent.

We will say that a linearly topologized module M is topologically artinian (resp. topologically noetherian, resp. topologically of finite length, resp. topologically coperfect) if M has a fundamental system $\{U\}$ of submodule neighbourhoods of 0, such that every module M/U is artinian (resp. noetherian, resp. of finite length, resp. coperfect) [1].

THEOREM 12.- Let $A = \begin{pmatrix} A_o & M \\ 0 & A_1 \end{pmatrix}$ be a (left) linearly topologized ring. Then A is (left) topologically artinian (resp. topologically noetherian, resp. topologically of finite length) if and only if A_1 and $A_o \amalg M$ are so for their natural (left) linear topologies.

PROOF: Let $\mathfrak{a}$ be a left open ideal of A. Then $\mathfrak{a} = \begin{pmatrix} & V_{\mathfrak{a}} \\ 0 & \mathfrak{a}_1 \end{pmatrix}$ $(M.\mathfrak{a}_1 \subset V_{\mathfrak{a}})$: where $\mathfrak{a}_1$ and $V_{\mathfrak{a}}$ are open submodules of A_1 and $A_o \amalg M$ respectively. Then by (25) above we have an exact sequence of A-modules

$$(27) \qquad 0 \longrightarrow (A_o \amalg M)/V_{\mathfrak{a}} \longrightarrow A/\mathfrak{a} \longrightarrow A_1/\mathfrak{a}_1 \longrightarrow 0$$

where A operates through A_o on $(A_o \amalg M)/V_{\mathfrak{a}}$ and through A_1 on $A_1/\mathfrak{a}_1$. Since in an exact sequence (in any abelian category)

$$0 \longrightarrow C_1 \longrightarrow C_2 \longrightarrow C_3 \longrightarrow 0$$

C_2 is noetherian (resp. artinian, resp. of finite length) if and only if C_1 and C_3 are noetherian (resp. ...), it follows easily from (27) that $A/\mathfrak{a}$ is artinian (resp. noetherian, resp. of finite length) as a left A-module if and only if $(A_o \amalg M)/V_{\mathfrak{a}}$ is artinian (resp. ...) as a left A_o-module and $A_1/\mathfrak{a}_1$ is artinian (resp. ...) as a left A_1-module, and this proves the Theorem 12.

COROLLARY 1.- A split linearly topologized ring $\begin{pmatrix} A_o & M \\ 0 & A_1 \end{pmatrix}$ is (left) topologically artinian (resp. topologically noetherian, resp. topologically of finite length), if and only if A_o, A_1 and M are so for their natural

[1] These conditions are then verified for all open submodules U.

(left) linear topologies.

In the discrete case we obtain:

COROLLARY 2.- A rings $A = \begin{pmatrix} A_o & M \\ 0 & A_1 \end{pmatrix}$ is left artinian (resp. left noetherian) if and only if the ring A_o, A_1 are left artinian (resp. left noetherian) and M is an artinian (resp. a finitely generated) left A_o-module [1].

COROLLARY 3.- Consider a stable extension of locally noetherian categories

$$0 \longrightarrow \underline{D} \xrightarrow{i_*} \underline{C} \xrightarrow{j^*} \underline{C}/\underline{D} \longrightarrow 0$$

Suppose that the endomorphism ring of a big injective $I_{\underline{D}}$ in $\underline{D}$ and $I_{\underline{C}/\underline{D}}$ in $\underline{C}/\underline{D}$ is a Leptin ring [2] (i.e. is topologically artinian). Then the endomorphism ring of the big injective $i_*(I_{\underline{D}}) \amalg j_*(I_{\underline{C}/\underline{D}})$ in $\underline{C}$ is a Leptin ring if and only if for every noetherian subobject N of $j_*(I_{\underline{C}/\underline{D}})$ in $\underline{C}$, the left $\mathrm{Hom}_{\underline{D}}(I_{\underline{D}}, I_{\underline{D}})$-module $\mathrm{Hom}_{\underline{C}}(N, i_*(I_{\underline{D}}))$ is artinian.

Example 1.- Let $\mathrm{Tors}(\mathbb{Z})$ be the category of torsion abelian groups, and consider the exact sequence

$$0 \longrightarrow \mathrm{Tors}(\mathbb{Z}) \longrightarrow \mathrm{Mod}(\mathbb{Z}) \longrightarrow \mathrm{Mod}(\mathbb{Q}) \longrightarrow 0$$

Here $\mathrm{Tors}(\mathbb{Z})$ and $\mathrm{Mod}(\mathbb{Q})$ are locally finite, and have big injectives $\mathbb{Q}/\mathbb{Z}$ and $\mathbb{Q}$ respectively. The endomorphism rings of these are $\prod_{p \in P} \mathbb{Z}_p$, [where P is the set of prime numbers, and $\mathbb{Z}_p$ is the ring of p-adic integers] and $\mathbb{Q}$ respectively. Both these rings are of course Leptin rings. But the endomorphism ring of $\mathbb{Q} \amalg \mathbb{Q}/\mathbb{Z}$ in $\mathrm{Mod}(\mathbb{Z})$ is not a Leptin ring, for $\mathbb{Z}$ is a noetherian subobject of $\mathbb{Q}$, and $\mathrm{Hom}_{\mathbb{Z}}(\mathbb{Z}, \mathbb{Q}/\mathbb{Z}) = \mathbb{Q}/\mathbb{Z}$ is

---

1) According to Hopkins [16], a left artinian ring is left noetherian. If we combine this with Corollary 1, we get that if A_o, A_1 are left artinian rings and M an A_o-A_1-bimodule that is a left artinian A_o-module, then M is finitely generated as a left A_o-module!

2) If the endomorphism ring of one big injective in a locally noetherian category $\underline{F}$ is a Leptin ring, then the endomorphism ring of any injective (big or not) in $\underline{F}$ is a Leptin ring (cf. [38],[39]).

not an artinian $\mathrm{End}_{\mathbb{Z}}(\mathbb{Q}/\mathbb{Z})$-module, for it has the following strictly decreasing chain of $\mathrm{End}_{\mathbb{Z}}(\mathbb{Q}/\mathbb{Z})$-submodules:

$$\mathbb{Q}/\mathbb{Z} = \coprod_{p\in P} \mathbb{Z}_{p_\infty} \supset \coprod_{\substack{p\geq 3\\ p\in P}} \mathbb{Z}_{p_\infty} \supset \coprod_{\substack{p\geq 5\\ p\in P}} \mathbb{Z}_{p_\infty} \supset \coprod_{\substack{p\geq 7\\ p\in P}} \mathbb{Z}_{p_\infty} \supset \coprod_{\substack{p\geq 11\\ p\in P}} \mathbb{Z}_{p_\infty} \supset \cdots$$

Here $\mathbb{Z}_{p_\infty}$ is the indecomposable injective of type p in $\mathrm{Mod}(\mathbb{Z})$, cf. [23].

Example 2.- Let $\mathbb{Z}_{(p)}$ be the localized ring of $\mathbb{Z}$ at the prime number p. This ring is noetherian and of Krull dimension 1, and one verifies using Corollary 3, that here the endomorphism ring of a big injective I is a Leptin ring A. This ring A can not be topologically noetherian too, for then it would be pseudo-compact, which is impossible, since the Krull dimension of $\mathrm{Mod}(\mathbb{Z}_{(p)})$ is 1. Thus $\mathrm{Dis}(A)$ is a locally artinian (and also locally coherent) category that is not locally noetherian. The theorem of Hopkins [16] cited above can be formulated as saying that if R is a ring, then $\mathrm{Mod}(R)$ locally artinian implies that $\mathrm{Mod}(R)$ is locally noetherian. The example $\mathrm{Dis}(A)$ shows that this theorem can not be extended to Grothendieck categories, at least not in this formulation.

THEOREM 13.- The (left) linearly topologized ring $A = \begin{pmatrix} A_0 & M \\ 0 & A_1 \end{pmatrix}$ is topologically coperfect, if and only if $A_0 \amalg M$ and A_1 are so.

START OF THE PROOF: If $\mathfrak{A} = \begin{pmatrix} & V_{\mathfrak{A}} \\ 0 & \mathfrak{A}_1 \end{pmatrix}$ $(M.\mathfrak{A}_1 \subset V_{\mathfrak{A}})$ is an open left ideal of A, then the exact sequence (27) above shows that if A is topologically coperfect, then $A_0 \amalg M$ and A_1 are so too. To prove the converse we need the following generalization of the Corollary of Proposition 13:

LEMMA 1.- With the notations above, the finitely generated left A-submodules of $\begin{pmatrix} (A_0 \amalg M)/V_{\mathfrak{A}} \\ 0 \quad A_1/\mathfrak{A}_1 \end{pmatrix}$ are exactly the A-submodules of the form $\begin{pmatrix} T + M.\mathfrak{b} \\ 0 \quad \mathfrak{b} \end{pmatrix}$, where $T \subset (A_0 \amalg M)/V_{\mathfrak{A}}$ is an arbitrary finitely generated A_0-submodule, and where $\mathfrak{b} \subset A_1/\mathfrak{A}_1$ is an arbitrary finitely generated

A_1-submodule ($M.\mathfrak{b}$ is defined by the natural product mapping $M \times A_1/\mathfrak{a}_1 \xrightarrow{\cdot} (A_0 \amalg M)/V_{\mathfrak{a}}$, which is well-defined, since $M.\mathfrak{a}_1 \subset V_{\mathfrak{a}}$).

PROOF: Direct computation.

END OF THE PROOF OF THEOREM 13: By Lemma 1, a decreasing sequence of finitely generated submodules of $A/\mathfrak{a}$ is necessarily of the form:

$$(28) \qquad \begin{pmatrix} T_1 + M.\mathfrak{b}_1 \\ 0 \quad \mathfrak{b}_1 \end{pmatrix} \supset \begin{pmatrix} T_2 + M.\mathfrak{b}_2 \\ 0 \quad \mathfrak{b}_2 \end{pmatrix} \supset \dots \supset \begin{pmatrix} T_n + M.\mathfrak{b}_n \\ 0 \quad \mathfrak{b}_n \end{pmatrix} \supset \dots \supset$$

where

$$(29) \qquad A_1/\mathfrak{a}_1 \supset \mathfrak{b}_1 \supset \mathfrak{b}_2 \supset \dots \supset \mathfrak{b}_n \supset \dots \supset$$

is a decreasing sequence of finitely generated A_1-submodules and where the T_i are finitely generated A_0-submodules of $(A_0 \amalg M)/V_{\mathfrak{a}}$ that however do not necessarily form a decreasing sequence. Since A_1 is supposed to be topologically coperfect, the sequence (29)must become stationary from a certain index n^* on, and so in particular we get from (28) a decreasing sequence

$$(30) \qquad T_{n^*} + M.\mathfrak{b}_{n^*} \supset T_{n^*+1} + M.\mathfrak{b}_{n^*} \supset T_{n^*+2} + M.\mathfrak{b}_{n^*} \supset \dots \supset$$

Now let $\left\{ f^{\,i}_{n^*+1} \right\}_{i=1}^{t}$ be an A_0-basis for $T_{n^*+1} \subset (A_0 \amalg M)V_{\mathfrak{a}}$. Then every $f^{\,i}_{n^*+1} = t^i_{n^*} + \varphi_i$, where $\varphi_i \in M.\mathfrak{b}_{n^*}$, and $t^i_{n^*} \in T_{n^*}$. Now these $t^i_{n^*}$ generate a (finitely generated!) submodule of T_{n^*}, which we will denote by T'_{n^*+1} and clearly $T'_{n^*+1} + M.\mathfrak{b}_{n^*} = T_{n^*+1} + M.\mathfrak{b}_{n^*}$. Continuing in this manner, we see that the decreasing sequence (30) can be written

$$(31) \qquad T_{n^*} + M.\mathfrak{b}_{n^*} \supset T'_{n^*+1} + M.\mathfrak{b}_{n^*} \supset T'_{n^*+2} + M.\mathfrak{b}_{n^*} \supset \dots$$

where

$$(32) \qquad (A_0 \amalg M)/V_{\mathfrak{a}} \supset T_{n^*} \supset T'_{n^*+1} \supset T'_{n^*+2} \supset \dots \supset$$

is a decreasing sequence of finitely generated submodules of $(A_0 \amalg M)V_{\mathfrak{a}}$. But $(A_0 \amalg M)/V_{\mathfrak{a}}$ is coperfect [$A_0 \amalg M$ is topologically coperfect]. It follows that (32) must become stationary and thus also (31) is stationary. Since (31) is just another way of writing (30), we finally see that (28) is stationary and so the Theorem 13 is proved.

COROLLARY 1.- *If* $A = \begin{pmatrix} A_0 & M \\ 0 & A_1 \end{pmatrix}$ *is linearly topologized with a split topology, then* A *is topologically coperfect if and only if* A_0 *and* A_1 *are so for their natural linear topologies.*

We have in fact that if A_0 is topologically coperfect, then every object of $\mathrm{Dis}(A_0)$ is coperfect.

COROLLARY 2.- *The ring* $\begin{pmatrix} A_0 & M \\ 0 & A_1 \end{pmatrix}$ *is right perfect if and only if* A_0 *and* A_1 *are so.*

PROBLEM: Is it true that the topology on a linearly topologized, topologically coperfect, complete ring $\begin{pmatrix} A_0 & M \\ 0 & A_1 \end{pmatrix}$ is a split topology?

We will not pass to the topologically coherent case, which by far is the most complicated one. For our purposes (cf. Theorem 11) it is sufficient to determine the necessary and sufficient conditions for $A = \begin{pmatrix} A_0 & M \\ 0 & A_1 \end{pmatrix}$ to be topologically coherent for a linear topology such that A has a fundamental system $\{\mathfrak{a}\}$ of *split* open ideal neighbourhoods of 0, such that every $A/\mathfrak{a}$ is coherent in $\mathrm{Dis}(A)$. We will say that A is ss-topologically coherent if this is true [we do not know if this is the same as saying that the linear topology on A is split and that A is topologically coherent (this could be called s-topological coherence)].

In order to express our results in a convenient form we will first need some generalities about linearly topologized (bi-)modules and their tensor products.

PROPOSITION 15.- *Let* A_0 *and* A_1 *be rings and let* M *be an* A_0-A_1-*bimodule. Suppose that* A_0 *and* M *are left linearly topologized and that* M *is a topological* A_0-A_1-*bimodule when* A_1 *is given the discrete topology*[1]. *Let* F *be a finitely generated left* A_1 *-module, and let* $\{f_i\}_1^t$ *be a fixed finite set of generators for* F. *Consider the* A_o*-submodules of* $M \otimes_{A_1} F$ *of the form*

$$V = V_U^{f_1, \ldots, f_t} = \operatorname{Im}\Big(\coprod_1^t U \longrightarrow \coprod_1^t M \xrightarrow{\sum_1^t (\ ) \otimes_{A_1} f_i} M \otimes_{A_1} F\Big)$$

where U *is an open* A_o*-submodule of* M. *These submodules* V *define a left* A_o *-linear topology on* $M \otimes_{A_1} F$, *that is independent of the choice of the finite set of generators* $\{f_i\}_1^t$ *of* F. *Furthermore* $M \otimes_{A_1} F$ *is a topological left* A_o *-module and for every* A_1 *-linear map* $F' \xrightarrow{\phi} F$ (F' *of finite type*), $Id_M \otimes_{A_1} \phi : M \otimes_{A_1} F' \longrightarrow M \otimes_{A_1} F$ *is a continuous* A_o*-linear map.*

PROOF: Let $\{g_i\}_1^s$ be another set of generators for F. I claim that every $V_U^{g_1, \ldots, g_s}$ contains a submodule of the form $V_U^{f_1, \ldots, f_t}$.

We have a factorization

$$(33) \qquad \begin{array}{ccccc} \coprod_1^t A_1 & \xrightarrow{\sum_1^t (\ )\cdot f_i} & F & \longrightarrow & 0 \\ \Big\downarrow \psi & & \Big\| & & \\ \coprod_1^s A_1 & \xrightarrow{\sum_1^s (\ )\cdot g_i} & F & \longrightarrow & 0 \end{array}$$

Here ψ must be given by right multiplication with a matrix (ψ_{ij}) of elements in A_1:

$$\{a_i\}_1^t \longmapsto \left\{ \sum_{i=1}^t a_i \psi_{ij} \right\}_{j=1}^s$$

---

1) This means that M is a left topological A-module, and that for any $a_1 \in A_1$ the right multiplication map $M \xrightarrow{\cdot a_1} M$ is continuous [this is true in particular if A_1 has some linear topology (not necessarily discrete) for which M is a topological right A_1-module].

By hypothesis there exists an open A_o-submodule $\tilde{U}$ of M such that $\tilde{U}\cdot\psi_{ij} \subset U$. Now if we combine this with (33) we get

$$V_{\tilde{U}}^{f_1, \dots, f_t} \subset V_U^{g_1, \dots, g_s} .$$

This shows that the linear topology on $M \otimes_{A_1} F$ is independent of the choice of the basis of F. The other results of the Proposition 15 are proved in the same way.

DEFINITION 5. -Let $\underline{C}$ be a Grothendieck category that is locally of finite type. We say that $C \in \underline{C}$ is pseudo-coherent (cf. Bourbaki [9], p. 62) if for every map $C' \xrightarrow{f} C$, where C' is of finite type, Ker f is so too. If A_o is a left linearly topologized ring, M a left linearly topologized A_o-module, then M is said to be topologically pseudo-coherent if M has a fundamental system U of A_o-submodule neighbourhoods of 0, such that every M/U is pseudo-coherent [in $\mathrm{Dis}(A_o)$]. We say that M is strongly topologically pseudo-coherent if in addition there is an open submodule W of M that is topologically coherent for the induced topology (then M/W is automatically pseudo-coherent, and it follows that there exists a fundamental system of such W:s).

We can now formulate the main result of this section:

THEOREM 14. -The following conditions of the left split linearly topologized ring $A = \begin{pmatrix} A_o & M \\ 0 & A_1 \end{pmatrix}$ are equivalent:

1) A is (left) ss-topologically coherent.

2) (i) The rings A_o and A_1 are topologically coherent for their natural left linear topologies;

(ii) For every coherent $C \in \mathrm{Dis}(A_1)$, $M \otimes_{A_1} C$ is strongly topologically pseudo-coherent for its natural topology, and for every monomorphism $C' \xrightarrow{\phi} C$ in $\mathrm{Coh}(\mathrm{Dis}(A_1))$, $(\mathrm{Id}_M \otimes_{A_1} \phi)^{-1}(V)$ is topologically coherent (induced topology) <u>if</u> $V \subset M \otimes_{A_1} C$ <u>is an open submodule that is topologically coherent</u> (it even suffices to require this for <u>one</u> such V).

<u>Remark</u> 1. -In the course of the proof of Theorem 14 we will see that (ii) can be replaced by the (apparently) weaker condition:

(ii)′ <u>There is a fundamental system</u> $\{\mathfrak{a}_1\}$ <u>of open left ideal neighbourhoods of</u> 0 <u>in</u> A_1, <u>such that:</u>

a) $A_1/\mathfrak{a}_1$ <u>is coherent in</u> $Dis(A_1)$;

b) $M \otimes_{A_1} A_1/\mathfrak{a}_1$ <u>is strongly topologically pseudo-coherent;</u>

c) <u>For all finitely generated</u> F_1, $F_1 \hookrightarrow A_1/\mathfrak{a}_1$, $M \otimes_{A_1} \frac{A_1/\mathfrak{a}_1}{F_1}$ <u>is topologically pseudo-coherent, and</u> $(Id_M \otimes_{A_1} i)^{-1}(V)$ <u>is topologically of finite type for at least one open submodule</u> $V \subset M \otimes_{A_1} A_1/\mathfrak{a}_1$ <u>that is topologically coherent.</u>

<u>Remark</u> 2. -In the discrete case Theorem 14 has a nice formulation in terms of Tor:

COROLLARY 1. -<u>The following conditions on the arbitrary generalized triangular matrix ring</u> $A = \begin{pmatrix} A_0 & M \\ 0 & A_1 \end{pmatrix}$ <u>are equivalent:</u>

1) A <u>is left coherent.</u>

2) (i) A_0 <u>and</u> A_1 <u>are left coherent rings;</u>

(ii) For every coherent $C \in Mod(A_1)$, <u>the left</u> A_0<u>-module</u> $Tor_1^{A_1}(M, C)$ <u>is pseudo-coherent for</u> $i = 0$ <u>and coherent for</u> $i \geq 1$.

Here we also have an (apparently) weaker formulation of (ii):

(ii)′$_{Cor. 1}$ <u>For all finitely generated left ideals</u> $\mathfrak{a}_1$ of A_1, $M \otimes_{A_1} A_1/\mathfrak{a}_1$ <u>is left</u> A_0 <u>-pseudo-coherent, and</u> $Tor_1^{A_1}(M, A_1/\mathfrak{a}_1)$ <u>is of finite type as a left</u> A_0 <u>-module.</u>

Remark 3. -There should also exist a formulation of the part (ii) of condition 2) of Theorem 14 in terms of linearly topologized Tor:s, a formulation that reduces to (ii) in Corollary 1 in the discrete case. We hope to return to this later.

Remark 4. -If A_1 is discrete, then Theorem 14 is simpler, and if A_1 is even a skew-field then we have:

COROLLARY 2. -*The following conditions on the left split linearly topologized ring* $A = \begin{pmatrix} A_o & M \\ 0 & A_1 \end{pmatrix}$ (A_1 = skew-field) *are equivalent:*

1) A *is left ss-topologically coherent.*

2) A_o *is left topologically coherent, and* M *is strongly pseudo-coherent as a left topological* A_o *-module.*

PROOF OF THEOREM 14: Suppose that $A = \begin{pmatrix} A_o & M \\ 0 & A_1 \end{pmatrix}$ has a split linear topology, let $\mathfrak{A} = \begin{pmatrix} \mathfrak{A}_o & U \\ 0 & \mathfrak{A}_1 \end{pmatrix}$ ($M\mathfrak{A}_1 \subset U$) be an open split left ideal of A and let us try to analyse the condition that $A/\mathfrak{A}$ is coherent in Dis(A). We have that

$$(34) \qquad A/\mathfrak{A} = \begin{pmatrix} A_o/\mathfrak{A}_o & M/U \\ 0 & A_1/\mathfrak{A}_1 \end{pmatrix}$$

where A operates to the left by matrix multiplication $[M \times A_1/\mathfrak{A}_1 \xrightarrow{\cdot} M/U$ is well-defined since $M\mathfrak{A}_1 \subset U]$. It is clear that (34) gives rise to a direct sum decomposition in Dis(A):

$$(35) \qquad A/\mathfrak{A} = A_o/\mathfrak{A}_o \amalg \begin{pmatrix} 0 & M/U \\ 0 & A_1/\mathfrak{A}_1 \end{pmatrix}$$

where A operates through A_o on $A_o/\mathfrak{A}_o$ and by left matrix multiplication on the right factor of (35). Thus:

$$(36) \qquad A/\mathfrak{A} \text{ coherent in Dis(A)} \iff \begin{cases} \alpha) & A_o/\mathfrak{A}_o \text{ is coherent in Dis(A).} \\ \beta) & \begin{pmatrix} 0 & M/U \\ 0 & A_1/\mathfrak{A}_1 \end{pmatrix} \text{ is coherent in Dis(A).} \end{cases}$$

It is clear that α) is equivalent to: For each open split left ideal $\tilde{\mathfrak{a}} = \begin{pmatrix} \tilde{\mathfrak{a}}_0 & \tilde{U} \\ 0 & \tilde{\mathfrak{a}}_1 \end{pmatrix} (M\tilde{\mathfrak{a}}_1 \subset \tilde{U})$ of A and for each left A-module map Φ (t arbitrary finite):

$$\coprod_1^t A/\tilde{\mathfrak{a}} = \coprod_1^t \begin{pmatrix} A_0/\tilde{\mathfrak{a}}_0 & M/\tilde{U} \\ 0 & A_1/\tilde{\mathfrak{a}}_1 \end{pmatrix} \xrightarrow{\Phi} A_0/\mathfrak{a}_0$$

Ker Φ is of finite type. But to give Φ is the same as to give elements $\{\bar{a}_0^i\}_1^t$ of $A_0/\mathfrak{a}_0$ such that $\tilde{\mathfrak{a}}_0 \cdot \bar{a}_0^i = 0 (i = 1, \ldots, t)$ in $A_0/\mathfrak{a}_0$: The multiplication map $A_0 \xrightarrow{\cdot \bar{a}_0^i} A_0/\mathfrak{a}_0$ passes then to the quotient

$$A_0/\tilde{\mathfrak{a}}_0 \xrightarrow{\cdot \bar{a}_0^i} A_0/\mathfrak{a}_0$$

and Φ is then given by:

$$\coprod_1^t \begin{pmatrix} A_0/\tilde{\mathfrak{a}}_0 & M/\tilde{U} \\ 0 & A_1/\tilde{\mathfrak{a}}_1 \end{pmatrix} \ni \left\{ \begin{pmatrix} \check{a}_0^i & \check{m}^i \\ 0 & \check{a}_1^i \end{pmatrix} \right\}_{i=1}^t \longmapsto \sum_{i=1}^t \check{a}_0^i \cdot \bar{a}_0^i \in A_0/\mathfrak{a}_0$$

and so

$$(37) \quad \operatorname{Ker} \Phi = \begin{pmatrix} \operatorname{Ker}\left(\coprod_1^t A_0/\tilde{\mathfrak{a}}_0 \xrightarrow{\sum_1^t (\ )\cdot \bar{a}_0^i} A_0/\mathfrak{a}_0\right) & \coprod_1^t M/\tilde{U} \\ 0 & \coprod_1^t A_1/\tilde{\mathfrak{a}}_1 \end{pmatrix}$$

which thus has to be of finite type.

Now let us turn to the condition β) of (36). It is clear that β) is equivalent to: For each open split left ideal $\tilde{\mathfrak{a}}$ of A and for each left A-module map (t arbitrary finite):

$$\coprod_1^t \begin{pmatrix} A_o/\tilde{\alpha}_o & M/\tilde{U} \\ 0 & A_1/\tilde{\alpha}_1 \end{pmatrix} \xrightarrow{\Psi} \begin{pmatrix} 0 & M/U \\ 0 & A_1/\alpha_1 \end{pmatrix}$$

Ker Ψ is of finite type. But to give Ψ is the same as to give elements $\left\{\begin{pmatrix} 0 & \bar{m}^i \\ 0 & \bar{a}_1^i \end{pmatrix}\right\}_{i=1}^t$ of $\begin{pmatrix} 0 & M/U \\ 0 & A_1/\alpha_1 \end{pmatrix}$ such that $\begin{pmatrix} \tilde{\alpha}_o & \tilde{U} \\ 0 & \tilde{\alpha}_1 \end{pmatrix} \cdot \begin{pmatrix} 0 & \bar{m}^i \\ 0 & \bar{a}_1^i \end{pmatrix} = 0$ $(i = 1, \ldots, t)$: These t relations can be written:

(38) $\tilde{\alpha}_o \cdot \bar{m}^i = 0$ (in M/U), $\tilde{U} \cdot \bar{a}_1^i = 0$ (in M/U), $\tilde{\alpha}_1 \cdot \bar{a}_1^i = 0$ (in A_1/α_1)

$(i = 1, \ldots, t)$

[the middle $\cdot$ is defined by restriction of the map $M \times A_1/\alpha_1 \to M/U$] and (38) can be expressed by saying that the natural multiplication maps

$$(39)\quad \left.\begin{array}{l} A_o \xrightarrow{\cdot \bar{m}^i} M/U \\ M \xrightarrow{\cdot \bar{a}_1^i} M/U \\ A_1 \xrightarrow{\cdot \bar{a}_1^i} A_1/\alpha_1 \end{array}\right\} \text{ define quotient maps } \left\{\begin{array}{l} A_o/\tilde{\alpha}_o \xrightarrow{\cdot \bar{m}^i} M/U \\ M/\tilde{U} \xrightarrow{\cdot \bar{a}_1^i} M/U \\ A_1/\tilde{\alpha}_1 \xrightarrow{\cdot \bar{a}_1^i} A_1/\alpha_1 \end{array}\right.$$

$(i = 1, \ldots, t)$

and Ψ is then given by [we use the notations in the formulas to the right of (39)]:

$$\coprod_1^t \begin{pmatrix} A_o/\tilde{\alpha}_o & M/\tilde{U} \\ 0 & A_1/\tilde{\alpha}_1 \end{pmatrix} \ni \left\{\begin{pmatrix} \check{a}_o^i & \check{m}^i \\ 0 & \check{a}_1^i \end{pmatrix}\right\}_{i=1}^t \longmapsto \begin{pmatrix} 0 & \sum_1^t (\check{a}_o^i \cdot \bar{m}^i + \check{m}^i \cdot \bar{a}_1^i) \\ 0 & \sum_1^t \check{a}_1^i \cdot \bar{a}_1^i \end{pmatrix}$$

and so

$$(40)\ \operatorname{Ker}\Psi = \begin{pmatrix} \operatorname{Ker}\left[\coprod_1^t (A_o/\tilde{\mathfrak{a}}_o \amalg M/\tilde{U}) \xrightarrow{\sum_{i=1}^t [(\ )\cdot\bar{m}^i + (\ )\cdot\bar{a}_1^i]} M/U\right] \\ 0 \qquad \operatorname{Ker}(\coprod_1^t A_1/\tilde{\mathfrak{a}}_1 \xrightarrow{\sum_{i=1}^t (\ )\cdot\bar{a}_1^i} A_1/\mathfrak{a}_1) \end{pmatrix}$$

which thus has to be of finite type.

In order to analyse (37) and (40), we now need the following general lemma, which generalizes Lemma 1 above.

LEMMA 2. -*Let* $A = \begin{pmatrix} A_o & M \\ 0 & A_1 \end{pmatrix}$ *be an arbitrary generalized triangular matrix ring and* $\tilde{\mathfrak{a}} = \begin{pmatrix} \tilde{V} & \\ 0 & \tilde{\mathfrak{a}}_1 \end{pmatrix}$ $(M\tilde{\mathfrak{a}}_1 \subset V)$ *an arbitrary left ideal of* A. *Then the finitely generated* A-*submodules of* $\coprod_1^t \begin{pmatrix} (A_o \amalg M)/\tilde{V} \\ 0 \quad A_1/\tilde{\mathfrak{a}}_1 \end{pmatrix}$ *are exactly the* A -*submodules of the form:*

$$\begin{pmatrix} \tilde{F}_o + \operatorname{Im}(M \otimes_{A_1} \tilde{F}_1 \xrightarrow{\cdot} \coprod_1^t (A_o \amalg M)/\tilde{V}) \\ 0 \qquad\qquad \tilde{F}_1 \end{pmatrix}$$

where $\tilde{F}_o \subset \coprod_1^t (A_o \amalg M)/\tilde{V}$ *is an arbitrary* A_o-*submodule of finite type, and* $\tilde{F}_1 \subset \coprod_1^t A_1/\tilde{\mathfrak{a}}_1$ *is an arbitrary* A_1-*submodule of finite type* [*here* $\cdot$ *is defined by* $M \times \coprod_1^t A_1/\mathfrak{a}_1 \xrightarrow{\cdot} \coprod_1^t (A_o \amalg M)/\tilde{V}$ $(M\cdot\tilde{\mathfrak{a}}_1 \subset \tilde{V})$].

PROOF: Direct computation.

Now the Lemma 2 implies that to say that $\operatorname{Ker}\Phi$ of (37) is of finite type is the same as to say that

$$\mathrm{Ker}\Big(\coprod_{1}^{t} A_o/\tilde{\mathfrak{a}}_o \xrightarrow{\sum_1^t (\ )\cdot\bar{a}_o^i} A_o/\mathfrak{a}_o\Big)$$

is of finite type as a left A_o-module.

Thus α) [cf.(36)] is equivalent to: $A_o/\mathfrak{a}_o$ is coherent in $\mathrm{Dis}(A_o)$.

The same Lemma 2 implies that $\mathrm{Ker}\ \Psi$ of (40) is of finite type if and only if:

A) $\tilde{F}_1 = \mathrm{Ker}\big(\coprod_1^t A_1/\tilde{\mathfrak{a}}_1 \xrightarrow{\sum_1^t (\ )\cdot\bar{a}_1^i} A_1/\mathfrak{a}_1\big)$ is of finite type as a left A_1-module.

B) $$\frac{\mathrm{Ker}\Big[\coprod_1^t (A_o/\tilde{\mathfrak{a}}_o \amalg M/\tilde{\mathfrak{V}}) \xrightarrow{\sum_1^t [(\ )\cdot\bar{m}_1^i + (\ )\cdot\bar{a}_1^i]} M/U\Big]}{\mathrm{Im}\big(M \otimes_{A_1} \tilde{F}_1 \xrightarrow{\ \cdot\ } \coprod_1^t M/\tilde{\mathfrak{V}}\big)}$$ is of finite type as a left A_1-module.

The universal validity of A) is clearly equivalent to: $A_1/\mathfrak{a}_1$ is coherent in $\mathrm{Dis}(A_1)$, but the analysis of B) is much more difficult: Put

$$F_1 = \mathrm{Im}\Big(\coprod_1^t A_1/\tilde{\mathfrak{a}}_1 \xrightarrow{\sum_1^t (\ )\cdot\bar{a}_1^i} A_1/\mathfrak{a}_1\Big).$$

Then the exact sequence

$$0 \longrightarrow \tilde{F}_1 \longrightarrow \coprod_1^t A_1/\tilde{\mathfrak{a}}_1 \longrightarrow F_1 \longrightarrow 0$$

gives rise to an exact sequence

$$(41)\qquad M \otimes_{A_1} \tilde{F}_1 \longrightarrow M \otimes_{A_1} \coprod_1^t A_1/\tilde{\mathfrak{a}}_1 \longrightarrow M \otimes_{A_1} F_1 \longrightarrow 0\ .$$

Consider the commutative exact diagram (R_o, R_1 and f are defined by the diagram):

(42)
$$\begin{array}{ccccccc}
 & & 0 & & 0 & & \\
 & & \uparrow & & \uparrow & & \\
 & & R_o/(0,\tilde{p}|)(R_1) & \xrightarrow{f} & \coprod_1^t A_o/\tilde{\mathfrak{a}}_o & & \\
 & & \uparrow & & \uparrow & & \\
0 & \longrightarrow & R_o & \longrightarrow & \coprod_1^t A_o/\tilde{\mathfrak{a}}_o \amalg \coprod_1^t M/\tilde{U} & \xrightarrow{\sum_1^t\left[(\ )\cdot\bar{m}^i + (\ )\cdot\bar{a}_1^i\right]} & M/U \\
 & & \uparrow{\scriptstyle (0,\tilde{p}|)} & & \uparrow{\scriptstyle (0,\tilde{p})} & & \uparrow{\scriptstyle p} \\
0 & \longrightarrow & R_1 & \longrightarrow & M \otimes_{A_1} \coprod_1^t A_1/\tilde{\mathfrak{a}}_1 & \xrightarrow{Id_M \otimes_{A_1} \sum_1^t (\ )\cdot\bar{a}_1^i} & M \otimes_{A_1} F_1
\end{array}$$

Here p is defined by restriction of the map $M \times A_1/\mathfrak{a}_1 \xrightarrow{\cdot} M/U$, $\tilde{p}$ is defined by the onto map $M \times A_1/\tilde{\mathfrak{a}}_1 \xrightarrow{\cdot} M/\tilde{U}(M\tilde{\mathfrak{a}}_1 \subset \tilde{U})$, and $\tilde{p}|$ is the restriction of $\tilde{p}$. If we compare (41) and (42) we see that the quotient of B) is the same as $R_o/(0,\tilde{p}|)(R_1)$. The analysis of α) above shows in particular that A ss-topologically coherent => A_o topologically coherent for its natural topology. Thus since we are interested in proving the equivalence of 1) and 2) in Theorem 14 we can and will suppose from now on that A_o is topologically coherent for its natural topology, and it is then permissible to suppose that $A_o/\tilde{\mathfrak{a}}_o$ is coherent in Dis(A_o). Under this hypothesis $\coprod_1^t A_o/\tilde{\mathfrak{a}}_o$ is coherent in Dis(A_o) and so $R_o/(0,\tilde{p}|)(R_1)$ is of finite type as an A_o-module if and only if Im f and Ker f [cf. (42)] are of finite type. But it follows from (42) that

(43)
$$\text{Im } f = \text{Ker}\left(\coprod_1^t A_o/\tilde{\mathfrak{a}}_o \xrightarrow{\sum_1^t (\ )\cdot\bar{m}^i} \frac{M/U}{\text{Im}\left(\coprod_1^t M/\tilde{U} \xrightarrow{\sum_1^t (\ )\cdot\bar{a}_1^i} M/U\right)}\right)$$

and that

(44)
$$\text{Ker } f = \frac{\text{Ker}(\coprod_1^t M/\tilde{U} \xrightarrow{\sum_1^t (\ )\cdot\bar{a}_1^i} M/U)}{\tilde{p}(R_1)} \quad .$$

Thus using the topology of Proposition 15 it follows from (43) that the universal validity of: Im f finitely generated is equivalent to:

(45) For all finitely generated submodules $F_1 \overset{i}{\hookrightarrow} A_1/\mathfrak{a}_1$, the quotient of $M \otimes_{A_1} \dfrac{A_1/\mathfrak{a}_1}{F_1}$ by the natural (open) image of U $(u \mapsto u \otimes \bar{1})$ is pseudo-coherent. To analyse (44), let us consider the commutative exact diagram

$$\begin{array}{ccccccccc}
0 & \longrightarrow & R_1 & \longrightarrow & M \otimes_{A_1} \coprod\limits_1^t A_1/\mathfrak{a}_1 & \xrightarrow{\;Id_M \otimes_{A_1} \sum\limits_1^t (\ )\cdot \bar{a}_1^i\;} & M \otimes_{A_1} F_1 & \longrightarrow & 0 \\
& & \downarrow {\scriptstyle \tilde{p}|} & & \downarrow {\scriptstyle \tilde{p}} & & \downarrow {\scriptstyle \check{p}} & & \\
0 & \longrightarrow & \mathrm{Ker}\Big(\coprod\limits_1^t M/\mathfrak{V} \xrightarrow{\sum\limits_1^t (\ )\cdot \bar{a}_1^i} M/U\Big) & \rightarrow & \coprod\limits_1^t M/U & \longrightarrow & \mathrm{Im}\Big(\coprod\limits_1^t M/\mathfrak{V} \xrightarrow{\sum\limits_1^t (\ )\cdot \bar{a}_1^i} M/U\Big) & \rightarrow & 0 \\
& & & & \downarrow & & \downarrow & & \\
& & & & 0 & & 0 & &
\end{array} \tag{46}$$

where $\check{p}$ is obtained by the natural factorization of p. An application of the snake lemma (|9|, Chap. 1, § 1, n° 4) to (46) gives an exact sequence

$$\mathrm{Ker}\ \tilde{p} \longrightarrow \mathrm{Ker}\ \check{p} \longrightarrow \mathrm{Coker}\ (\tilde{p}|) \longrightarrow 0$$

But $\mathrm{Coker}\ (\tilde{p}|) = \mathrm{Ker}\ f$ and $\mathrm{Ker}\ \tilde{p} = \coprod\limits_1^t \mathfrak{V}/M\tilde{\mathfrak{a}}_1$, so that to say that $\mathrm{Ker}\ f$ is of finite type is equivalent to say that

(47) $\left\{ \begin{array}{l} \mathrm{Coker}\left\{ \coprod\limits_1^t \mathfrak{V} \longrightarrow \mathrm{Ker}\Big[M \otimes_{A_1} F_1 \longrightarrow \mathrm{Im}\Big(\coprod\limits_1^t M/\mathfrak{V} \xrightarrow{\sum\limits_1^t (\ )\cdot \bar{a}_1^i} M/U\Big)\Big] \right\} \\ \text{is of finite type (the maps are the natural ones).} \end{array} \right.$

Note that in (47) F_1 is any finitely generated subobject of $A_1/\mathfrak{a}_1$, $\{\bar{a}_1^i\}_1^t$ is any finite set of generators for F_1, and $\mathfrak{V}$ is any open A_0-submodule of M such that $\mathfrak{V}\cdot\bar{a}_1^i = 0$ (in M/U)$(i=1, \ldots, t)$.

Consider now the exact sequence

$$(48) \qquad M \otimes_{A_1} F_1 \xrightarrow{Id_M \otimes_{A_1} i} M \otimes_{A_1} A_1/\mathfrak{a}_1 \longrightarrow M \otimes_{A_1} \frac{A_1/\mathfrak{a}_1}{F_1} \longrightarrow 0.$$

From (48) it follows that with the notations of Proposition 15, (47) can just be reformulated by saying that the open submodule $(Id_M \otimes_{A_1} i)^{-1}(V_U^{\overline{I}})$ of $M \otimes_{A_1} F_1$ is topologically of finite type. If we choose $F_1 = A_1/\mathfrak{a}_1$, we see in particular that $V_U^{\overline{I}}$ is topologically of finite type.

Now from all that has been proved it follows easily that with the notations and hypotheses of Theorem 14 we have the implication 1) => 2), where in 2) we have replaced (ii) by the weaker (ii)'. That (i) + (ii)' => 1) under the same hypotheses is also clear, and so we now have Theorem 14 with (ii) replaced by (ii)', and this implies in particular that Corollary 1 [with the weakened $(ii)'_{Cor.\ 1}$] and that Corollary 2 are valid.

Now the passage from (ii)' to (ii) [we suppose that (i) is satisfied] is an exercise in the use of topological pseudo-coherence and exact sequences, that we will only do here in the discrete case, where we will prove the more precise result required by Corollary 1: We suppose that $M \otimes_{A_1} A_1/\mathfrak{a}_1$ is pseudo-coherent and that $Tor_1^{A_1}(M, A_1/\mathfrak{a}_1)$ is of finite type for all left ideals $\mathfrak{a}_1 \subset A_1$ of finite type. Let us first prove by induction on the number of generators of C, that $M \otimes_{A_1} C$ is pseudo-coherent for all coherent C: if C has $\leq n$ generators, then we can find an exact sequence

$$0 \longrightarrow C_{n-1} \longrightarrow C \longrightarrow A_1/\mathfrak{a}_1 \longrightarrow 0$$

where $\mathfrak{a}_1$ is of finite type, and where C_{n-1} is coherent and has $\leq n-1$ generators. The exact sequence of Tor gives:

$$(49) \qquad Tor_1^{A_1}(M, A_1/\mathfrak{a}_1) \longrightarrow M \otimes_{A_1} C_{n-1} \longrightarrow M \otimes_{A_1} C \longrightarrow M \otimes_{A_1} A_1/\mathfrak{a}_1 \longrightarrow 0$$

and since the Tor_1 in (49) is of finite type, its image in $M \otimes_{A_1} C_{n-1}$ is so too, thus this image is coherent, since by the induction hypothesis $M \otimes_{A_1} C_{n-1}$ is pseudo-coherent. But then the quotient of $M \otimes_{A_1} C_{n-1}$ with this image is pseudo-coherent, and so we get from (49) an extension

$$0 \longrightarrow PC_1 \longrightarrow M \otimes_{A_1} C \longrightarrow M \otimes_{A_1} A_1/\mathfrak{a}_1 \longrightarrow 0$$

where the extreme terms are pseudo-coherent. Thus the middle term is so too. Now that we know that $M \otimes_{A_1} C$ is pseudo-coherent for all coherent C, the exact sequence

$$0 \longrightarrow \mathrm{Tor}_1^{A_1}(M, A_1/\mathfrak{a}_1) \longrightarrow M \otimes_{A_1} \mathfrak{a}_1 \longrightarrow M \longrightarrow M \otimes_{A_1} A_1/\mathfrak{a}_1 \longrightarrow 0$$

gives that $\mathrm{Tor}_1^{A_1}(M, A_1/\mathfrak{a}_1)$ is coherent, for it is a subobject of finite type of $M \otimes_{A_1} \mathfrak{a}_1$, which is pseudo-coherent, since the ideal $\mathfrak{a}_1$ is a coherent module (A_1 is coherent). From this we get, essentially as above by induction on the number of generators of C, first that $\mathrm{Tor}_1^{A_1}(M, C)$ is of finite type for all coherent C. Thus it follows in particular that $\mathrm{Tor}_2^{A_1}(M, A_1/\mathfrak{a}_1) \xrightarrow{\sim} \mathrm{Tor}_1^{A_1}(M, \mathfrak{a}_1)$ is of finite type, and using this we can go one step further and show that $\mathrm{Tor}_1^{A_1}(M, C)$ is coherent for all coherent C, and now one passes easily to the higher Tor:s and so the Corollary 1 is completely proved. As an immediate consequence we have:

COROLLARY 3. - If M is right A_1-flat and the ring A_o is left noetherian, then $\begin{pmatrix} A_o & M \\ 0 & A_1 \end{pmatrix}$ is left coherent if and only if A_1 is left coherent.

If we now confront the results of this § with those of § 5, then we obtain in particular:

THEOREM 15. -Let $\underline{C}_o$ and $\underline{C}_1$ be locally noetherian categories, A_o and A_1 the associated linearly topologized, topologically coherent, topologically coperfect, complete, sober rings. The map that to each left linearly topologized complete topological A_1-bimodule M [with $M \times A_1 \longrightarrow M$ left equicontinuous] that satisfies the condition (ii) of Theorem 14, associates the extension:

$$0 \longrightarrow TC(A_o)^o \longrightarrow TC\begin{pmatrix} A_o & M \\ 0 & A_1 \end{pmatrix}^o \longrightarrow TC(A_1)^o \longrightarrow 0$$

defines a one-one correspondence between the topological isomorphy classes of these bimodules M, and the equivalence classes of stable extensions

$$0 \longrightarrow \underline{C}_o \longrightarrow \underline{C} \longrightarrow \underline{C}_1 \longrightarrow 0$$

of locally noetherian categories.

Remark 1. -There are also other stable extensions, where the middle caregory is not locally noetherian.

Remark 2. -More details about Theorem 15 in the discrete case will be given in § 9 - § 10.

§ 9. Change of Krull dimension in stable extension of locally noetherian categories.

For simplicity we will only study the discrete case in this section. Thus let A_o and A_1 be left coherent, right perfect rings, and M a bimodule as in the Corollary 1 of Theorem 14 and consider the corresponding stable exact sequence

$$(50) \qquad 0 \longrightarrow TC(A_o)^o \longrightarrow TC\begin{pmatrix} A_o & M \\ 0 & A_1 \end{pmatrix}^o \longrightarrow TC(A_1)^o \longrightarrow 0.$$

Of course all categories in (50) have finite Krull dimension, and according to a general result of Gabriel [25], we have the inequality:

$$(51) \quad \max\left(\dim TC(A_o)^o, \dim TC(A_1)^o\right) \leq \dim TC(A)^o \leq \dim TC(A_o)^o + \dim TC(A_1)^o + 1,$$

where $A = \begin{pmatrix} A_o & M \\ 0 & A_1 \end{pmatrix}$. Thus the problem arises to formulate explicitly in terms of M the condition that the middle dimension in (42) attains a given value between the bounds of (51). Here we will only study the case when $\dim TC(A_1)^o = 0$, i.e. the case when A_1 is left artinian.

THEOREM 16. - Let A_o be a left coherent, right perfect ring, A_1 a left artinian ring, and M an A_o-A_1-bimodule satisfying the conditions [1] of Cor.1 of V Theor.14 and let A be as above. If M is not finitely generated as a left A_o-module [2], then $\dim TC(A)^o = \dim TC(A_1)^o + 1$, otherwise $\dim TC(A)^o = \dim TC(A_1)^o$. Further $TC(A_o)^o \subset TC(A)^o$ is exactly the last step of the Gabriel filtration of $TC(A)^o$ if and only if for all coherent $C \neq 0$ in $\mathrm{Mod}(A_1)$, $M \otimes_{A_1} C$ is not coherent (i.e. not of finite type, cf. [2]) in $\mathrm{Mod}(A_o)$.

The proof of this theorem is not difficult, and one can even prove that it is sufficient to suppose C cyclic in the last condition.

Example: If K is a field, then $\dim TC\begin{pmatrix} K & \coprod_{\aleph_0} K \\ 0 & K \end{pmatrix}^o = 1$, and using variations of this example, one can construct for any integer n, a left coherent, right perfect ring A, such that $\dim TC(A)^o = n$, and such that the (finite) Gabriel filtration of $TC(A)^o$ is stable under injective envelopes. Such rings A will be called stable (it is easy to formulate the stability condition purely ring-theoretically, using § 7) and they will be studied in detail in the next section.

§ 10. Application 1: The structure of right perfect, left coherent, stable rings.

To obtain what the title promises, we now only have to put together suitable pieces of § 5 and §§ 8-9. This gives:

THEOREM 17.- Let A be a left coherent, right perfect, stable ring. Then A can be built up in one and only one way as follows:

There exists a finite string of left artinian rings $A^o = A_o, A^1, \dots, A^n$, and an A_o-A^1-bimodule M^o such that $\mathrm{Tor}_i^{A^1}(M^o, \mathrm{Coh})$ is A_o-pseudo-coherent,

---

1) Note that by Hopkins [16] all left ideals of A_1 are finitely generated, and thus all finitely generated modules in $\mathrm{Mod}(A_1)$ are coherent.

2) Since M is left pseudo-coherent, M not of finite type is equivalent to: M is not left coherent.

but not coherent for Coh $\neq 0$ and $i = 0$, and coherent for $i \geq 1$. Put $A_1 = \begin{pmatrix} A_o & M^o \\ 0 & A^1 \end{pmatrix}$. There exists an A_1-A^2-bimodule M^1 such that $\mathrm{Tor}_i^{A^2}(M^1, \mathrm{Coh})$ is ..... etc.

Finally we arrive at an A_{n-1}-A^n-bimodule M^{n-1} satisfying the Tor-condition and such that $A = \begin{pmatrix} A_{n-1} & M^{n-1} \\ 0 & A^n \end{pmatrix}$. Furthermore $A^o, A^1, \ldots, A^n$ and $M^o, \ldots, M^{n-1}$ (and $A_o = A^o, A_1, \ldots, A_{n-1}$) are uniquely determined by A and conversely, if we give A^i:s and M^i:s satisfying the conditions above, and construct A as indicated, then A is left coherent, right perfect and stable and $\dim TC(A)^o = n$.

COROLLARY.- If A is a left coherent, right perfect, stable ring, then A is semi-primary, i.e. $\mathrm{rad}_J(A)$ is nilpotent, and $A/\mathrm{rad}_J(A)$ is a semi-simple artinian ring.

This corollary is proved by induction, using the fact that

$$\mathrm{rad}_J(A_t) = \begin{pmatrix} \mathrm{rad}_J A_{t-1} & M^{t-1} \\ 0 & \mathrm{rad}_J A^t \end{pmatrix}, \quad 1 \leq t \leq n \quad (A_n = A).$$

Remark 1.- We doubt that $\mathrm{rad}_J(A)$ is nilpotent if we leave out the stability condition in the corollary.

Remark 2.- Theorem 17 above implies that Mod(A) can be built up canonically as an iterated comma category of module categories over artinean rings. In fact, tensoring with the M^i defines functors $\mathrm{Mod}(A^{i+1}) \xrightarrow{T^i} \mathrm{Mod}(A_i)$ and it is very easy to see, that with the notations of [37] we have natural equivalences $(\mathrm{Mod}\ (A_i)\ ,\ T^i) \xrightarrow{\simeq} \mathrm{Mod}(A_{i+1})$, $0 \leq i \leq n-1$. Furthermore, one can verify that in this case, not only is $\mathrm{Mod}(A_i)$ a localizing subcategory of $\mathrm{Mod}(A_{i+1})$, but it is also a tri-localizing one, (i.e. the quotient functor has a left adjoint, which has a left adjoint). We hope to return to a systematic study of this. For the bilocalizing case, see [48]. The tetralocalizing subcategories etc. are not so interesting since they are direct factors.

§ 11. "Application" 2: Quasi-Frobenius categories.

Recall [16] that a ring A is said to be a (left) quasi-Frobenius ring if A is left artinian and left self-injective (then A is also right quasi-Frobenius). In [21] it was proved that if R is an arbitrary ring, then R is quasi-Frobenius if and only if every injective in Mod(.R) is projective. [The "dual" of this is also true, cf. [20].] Thus it is natural to introduce the following

DEFINITION 6.- *Let* $\underline{C}$ *be a Grothendieck category. We say that* $\underline{C}$ *is a quasi-Frobenius category, if every injective in* $\underline{C}$ *is also projective.*

Remark 1.- It is unnatural to use the "dual" result [20] as a basis for a definition, since there are non-zero Grothendieck categories (even satisfying AB 6) having no projective objects except 0 [49].

It was proved in [53] that a quasi-Frobenius category $\underline{C}$ has bounded decompositions of injectives. If furthermore $\underline{C}$ satisfies the axiom AB 6, then $\underline{C}$ is locally noetherian, and the injective envelope of every noetherian object in $\underline{C}$ is noetherian [53]. As we have said above, it follows from [21], that if $\underline{C}$ is also a module category Mod(R), then R is quasi-Frobenius. In the general case we do not even know if $\underline{C}$ locally noetherian + quasi-Frobenius => $\underline{C}$ locally finite, but we will now reformulate this in a purely ring-theoretical manner, and then prove that it is true in the special case when $\underline{C}$ has a big noetherian injective:

THEOREM 18.- *The following conditions on a linearly topologized, topologically coherent, topologically coperfect and complete ring* A *are equivalent:*

(i) $TC(A)^{o}$ *is a quasi-Frobenius category;*

(ii) A *is an injective object in* TC(A);

(iii) *For every topologically coherent closed left ideal* $\mathfrak{a}$ *of* A, *every continuous left* A-*linear map* $\mathfrak{a} \longrightarrow A$ *is the right multiplication by an element of* A, *and every* Ae (e primitive) *is artinian in* TC(A).

Remark.- We do not know if the last part of (iii) is a consequence of the first part.

PROOF: Follows from § 5 and from what has been said above.

PROBLEM: Let A satisfy the equivalent conditions of Theorem 18. Is it true that A is pseudo-compact? More precisely, if A is sober, is it true that A has a fundamental system of neighbourhoods of 0, formed by bilateral ideals $\mathfrak{a}_\alpha$ such that every $A/\mathfrak{a}_\alpha$ is a quasi-Frobenius ring [thus $A \xrightarrow{\sim} \varprojlim A/\mathfrak{a}_\alpha$.]? In this case, is the dual ring A^* the same as A^{opp}?

Suppose now that A is discrete. Then the ideals of condition (iii) of Theorem 18 are exactly the finitely generated (f.g.) left ideals, and (iii) can be expressed by saying that A is f.g. left self-injective. Now, both J.-E. Björk [7] and B. Stenström (unpublished) have proved independently that if A is a right perfect, left coherent and left self-injective ring, then A is a quasi-Frobenius ring, and the proof of Stenström has the merit that it extends immediately to the f.g. self-injective case:

THEOREM 19.- A right perfect, left coherent, left f.g. self-injective ring is a quasi-Frobenius ring (i.e. the problem above has a positive solution in the discrete case).

PROOF (Stenström): Let A be a ring with unit, S a subset of A. Put $l(S) = \{a \in A \mid aS = 0\}$ and $r(S) = \{a \in A \mid Sa = 0\}$. We need:

LEMMA 1: If A is left f.g. self-injective and left coherent, then $rl(I) = I$ for all finitely generated right ideals I of A.

PROOF: This is essentially a slight reformulation of what is proved in [16], p. 399-400.

LEMMA 2.- If A is left coherent, and I is a finitely generated right ideal, then $l(I)$ is a finitely generated left ideal.

PROOF: This follows from Chase [13], theorem 2.2.

LEMMA 3.- <u>If</u> A <u>is as in Theorem</u> 19, <u>then</u> A <u>is right noetherian</u>.

PROOF: It suffices to show that any ascending chain $\{I_n\}$ of finitely generated right ideals is stationary. By Lemma 2, $\{l(I_n)\}$ is a descending chain of finitely generated left ideals, and so it is stationary, since A is left coperfect. Now Lemma 1 completes the proof. To prove Theorem 19, it is now sufficient to remark, that a small modification of the proof of theorem 1 in [20] gives that a right noetherian, left f.g. self-injective ring is a quasi-Frobenius ring.

§ 12. <u>Final remarks</u>.

The biggest gap in the theory developed above is evidently the lack of a theory of general (non-stable) extensions of locally noetherian categories. To fill this gap, we would have to generalize the theory of generalized triangular matrix rings of § 8 to a theory of matrix rings of the form

$$A = \begin{pmatrix} A_o & M \\ N & A_1 \end{pmatrix},$$

where M and N are not only A_o-A_1- and A_1-A_o-bimodules respectively, but where there is also given an A_o-A_o-bimodule map $M \otimes_{A_1} N \xrightarrow{\varphi} A_o$ and an A_1-A_1-bimodule map $N \otimes_{A_o} M \xrightarrow{\psi} A_1$ such that the diagrams

$$\begin{array}{ccc} M \otimes_{A_1} (N \otimes_{A_o} M) & \xrightarrow{Id_M \otimes \psi} & M \otimes_{A_1} A_1 \\ \downarrow{\scriptstyle \varphi \otimes Id_M} & & \downarrow \wr \\ A_o \otimes_{A_o} M & \xrightarrow{\sim} & M \end{array} \qquad \begin{array}{ccc} (N \otimes_{A_o} M) \otimes_{A_1} N & \xrightarrow{Id_N \otimes \varphi} & N \otimes_{A_o} A_o \\ \downarrow{\scriptstyle \psi \otimes Id_M} & & \downarrow \wr \\ A_1 \otimes_{A_1} N & \xrightarrow{\sim} & N \end{array}$$

are commutative, and where multiplication in A is defined by matrix multiplication (using φ and ψ). This seems to be a formidable algebraic problem, and we might try to make an attack on this at a later occasion.

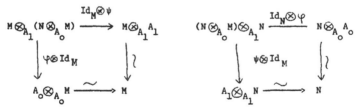

It also remains to make a detailed study of the associated topological rings in the examples 1)-7) of the introduction. The case 2) seems to be particularly rewarding, since any localizing subcategory of Qcoh(X) is stable, thus in particular the Gabriel filtration is stable under injective envelopes, and the bimodules that describe the corresponding extensions should be of interest. [We should make a duality theory for Qcoh(X) etc. ....] It should be remarked, that in the examples 5) and 7), parts of the associated topological rings have already been studied explicitly with nice applications in [11] [1], [29] and [24]. Finally, some structure theorems for some classes of strictly linearly compact rings, can be found in [1], [2], [18] and [32].

---

1) Brumer [11] uses a more restrictive definition of pseudo-compact rings than the usual one - he requires the existence of a fundamental system of open bilateral ideal neighbourhoods of 0 ...

BIBLIOGRAPHY

[1] I. KR. AMDAL - F. RINGDAL, Catégories unisérielles, Comptes rendus, série A, 267, 1968, p. 85 - 87 and p. 247 - 249.

[2] B. BALLET, Structure des anneaux strictement linéairement compacts commutatifs, Comptes rendus, série A, 266, 1968, p. 1113 - 1116.

[3] H. BASS, Finistic dimension and a homological generalization of semi-primary rings, Trans. Amer. Math. Soc., 95, 1960, p. 466 - 488.

[4] H. BASS, Injective dimension in noetherian rings, Trans. Amer. Math. Soc., 102, 1962, p. 18 - 29.

[5] H. BASS, Lectures on Topics in Algebraic K-theory, Tata Institute of Fundamental Research, Bombay, 1967.

[6] J.-E. BJÖRK, Rings satisfying a minimum condition on principal ideals (to appear in J. reine ang. Math.).

[7] J.-E. BJÖRK, On perfect rings, QF-rings and rings of endomorphisms of injective modules (to appear).

[8] N. BOURBAKI, Topologie générale, Chap. 1-2, 3e éd., Hermann, Paris, 1961.

[9] N. BOURBAKI, Algèbre commutative, Chap. 1-2, Hermann, Paris, 1961.

[10] N. BOURBAKI, Algèbre commutative, Chap. 3-4, Hermann, Paris, 1961.

[11] A. BRUMER, Pseudo-compact algebras, profinite groups and class formations, J. Algebra, 4, 1966, p. 442 - 470.

[12] H. CARTAN - S. EILENBERG, Homological Algebra, Princeton University Press, Princeton, 1956.

[13] S. U. CHASE, Direct products of modules, Trans. Amer. Math. Soc., 97, 1960, p. 457 - 473.

[14] S. U. CHASE, A generalization of the ring of triangular matrices, Nagoya Math. J., 18, 1961, p. 13 - 25.

[15] J. M. COHEN, Coherent graded rings and the non-existence of spaces of finite stable homotopy type (to appear in Comm. Math. Helv.).

[16] C. W. CURTIS - I. REINER, Representation Theory of Finite Groups and Associative Algebras, Wiley, New York, 1962.

[17] P. DELIGNE, Cohomologie à support propre et construction du foncteur $f^!$, Appendix to [34].

[18] J. DIEUDONNÉ, Topics in Local Algebra, Notre Dame Math. Lectures, n° 10, 1967.

[19] C. FAITH - Y. UTUMI, Quasi-injective modules and their endomorphism rings, Arch. der Math., 15, 1964, p. 166 - 174.

[20] C. FAITH, Rings with ascending condition on annihilators, Nagoya Math. J., 27, 1966, p. 179 - 191.

[21] C. FAITH - E. A. WALKER, Direct-sum representations of injective modules, J. Algebra, 5, 1967, p. 203 - 221.

[22] P. FREYD, Functor Categories and their Application to Relative Homology (mimeographed notes), Columbia University, 1962.

[23] P. GABRIEL, Objets injectifs dans les catégories abéliennes, Séminaire Dubreil-Pisot, 12, 1958 - 1959, Exposé 17.

[24] P. GABRIEL, Sur les catégories abéliennes localement noethériennes et leurs applications aux algèbres étudiées par Dieudonné, Séminaire J.-P. Serre, Collège de France, 1959 - 1960.

[25] P. GABRIEL, Des catégories abéliennes, Bull. Soc. Math. Fr., 90, 1962, p. 323 - 448.

[26] P. GABRIEL, Groupes formels, Séminaire Demazure-Grothendieck, I.H.E.S., 1963 - 1964, Fasc. 2 b, Exposé VII_B (cf. Math. Rev. 35, 1968, # 4222).

[27] P. GABRIEL - U. OBERST, Spektralkategorien und reguläre Ringe im von Neumannschen Sinn, Math. Z., 92, 1966, p. 389 - 395.

[28] P. GABRIEL - R. RENTSCHLER, Sur la dimension des anneaux et ensembles ordonnés, Comptes rendus, série A, 265, 1967, p. 712 - 715.

[29] E. S. GOLOD - I. R. ŠAFAREVIČ, O bašne polej klassov, Izv. Akad. Nauk SSSR, ser. mat., 28, 1964, p. 261 - 272. English translation in: American Mathematical Society Translations, Ser. 2, 48, 1965, p. 91 - 102.

[30] A. GROTHENDIECK, Sur quelques points d'algèbre homologique, Tohoku Math. J., 9, 1957, p. 119 - 221.

[31] A. GROTHENDIECK, Technique de descente et théorèmes d'existence en géométrie algébrique, II: Le théorème d'existence en théorie formelle des modules, Séminaire Bourbaki, 12, 1959 - 1960, Exposé 195. (Has also been published by Benjamin, New York, 1966.).

[32] A. GROTHENDIECK (and J. DIEUDONNÉ), Eléments de géométrie algébrique, Publ. Math. I.H.E.S., n° 20, 1964.

[33] M. HARADA, On a special type of hereditary abelian categories, Osaka J. Math., 4, 1967, p. 243 - 256.

[34] R. HARTSHORNE, Residues and Duality, Springer, Berlin, 1966.

[35] I. KAPLANSKY, Dual modules over a valuation ring, Proc. Amer. Math. Soc., 4, 1953, p. 213 - 219.

[36] I. KAPLANSKY, Infinite Abelian Groups, University of Michigan Press, Ann Arbor, 1954.

[37] F. W. LAWVERE, The category of categories as a foundation for mathematics, Proceedings of the Conference on Categorical Algebra, La Jolla 1965, Springer, Berlin, 1966, p. 1 - 20.

[38] H. LEPTIN, Linear kompakte Moduln und Ringe, I., Math. Z., 62, 1955, p. 241 - 267.

[39] H. LEPTIN, Linear kompakte Moduln und Ringe, II., Math. Z., 66, 1956 - 1957, p. 289 - 327.

[40] E. MATLIS, Injective modules over noetherian rings, Pacific J. Math., 8, 1958, p. 511 - 528.

[41] B. MITCHELL, Theory of Categories, Academic Press, New York, 1965.

[42] C. NĂSTĂSESCU - N. POPESCU, Quelques observations sur les topos abéliens, Rev. Roum. Math. Pures et Appl., 12, 1967, p. 553 - 563.

[43] Y. NOUAZÉ, Catégories localement de type fini et catégories localement noethériennes, Comptes rendus, 257, 1963, p. 823 - 824.

[44] F. OORT, Commutative Group Schemes, Springer, Berlin, 1966.

[45] Z. PAPP, On algebraically closed modules, Publ. Math. Debrecen, 6, 1959, p. 311 - 327.

[46] J.-E. ROOS, Sur les foncteurs dérivés de $\varprojlim$. Applications. Comptes rendus, 252, 1961, p. 3702 - 3704.

[47] J.-E. ROOS, Bidualité et structure des foncteurs dérivés de $\varprojlim$ dans la catégorie des modules sur un anneau régulier, Comptes rendus, 254, 1962, p. 1556 - 1558 and p. 1720 - 1722.

[48] J.-E. ROOS, Caractérisation des catégories qui sont quotients de catégories de modules par des sous-catégories bilocalisantes, Comptes rendus, 261, 1965, p. 4954 - 4957.

[49] J.-E. ROOS, Sur les foncteurs dérivés des produits infinis dans les catégories de Grothendieck. Exemples et contre-examples, Comptes rendus, 263, 1966, p. 895 - 898.

[50] J.-E. ROOS, Sur la condition AB 6 et ses variantes dans les catégories abéliennes, Comptes rendus, série A, 264, 1967, p. 991 - 994.

[51] J.-E. ROOS, Locally distributive spectral categories and strongly regular rings, Reports of the Midwest Category Seminar, Springer, Berlin, 1967, p. 156 - 181.

[52] J.-E. ROOS, Sur l'anneau maximal de fractions des AW^*-algèbres et des anneaux de Baer, Comptes rendus, série A, 266, 1968, p. 120 - 123.

[53] J.-E. ROOS, Sur la décomposition bornée des objets injectifs dans les catégories de Grothendieck, Comptes rendus, série A, 266, 1968, p. 449 - 452.

[54] J.-E. ROOS, Sur la structure des catégories abéliennes localement noethériennes, Comptes rendus, série A, 266, 1968, p. 701 - 704.

[55] Séminaire H. CARTAN, 11, 1958 - 1959, Exposé 15. (Has also been published by Benjamin, New York, 1967.)

[56] J.-P. SERRE, Groupes proalgébriques, Publ. Math. I.H.E.S. n° 7, 1960.

[57] J.-P. SERRE, Algèbre Locale. Multiplicités (rédigé par P. Gabriel), Springer, Berlin 1965.

[58] J.-P. SERRE, Cohomologie Galoisienne, Springer, Berlin, 1965.

[59] Y. UTUMI, On a theorem on modular lattices, Proc. Jap. Acad., 35, 1959, p. 16 - 21.

[60] J.-L. VERDIER, Topologies et faisceaux, Fasc. 1 of Séminaire Artin-Grothendieck, I.H.E.S., 1963 - 1964.

KAN EXTENSIONS, COTRIPLES AND ANDRÉ (CO)HOMOLOGY

by

Friedrich Ulmer[1]

INTRODUCTION

The aim of this note is to point out that the "non-abelian" derived functors A_* and H_* of André [1] and Barr-Beck [2] are the abelian derived functors of the Kan extension and to sketch how the properties of A_* and H_* can be obtained within the abelian framework. In many cases the abelian viewpoint leads to simplifications of proofs and generalizations of known facts as well as to new results and insights. An instance of the latter is the method of acyclic models which turns out to be the standard procedure in homological algebra for computing the left derived functors by means of projectives (cf. [20] this volume). We mostly limit ourselves to dealing with homology and leave it to the reader to state the dual theorems for cohomology. To make sure that this works in practice, we try to avoid exactness conditions on the range category (i.e. Grothendieck's axiom AB5), [12]).

---

[1] Part of this work was supported by the Forschungsinstitut für Mathematik der E.T.H. and the Deutsche Forschungsgemeinschaft.

However a few results depend on AB5) and are probably false without it.

Our approach is based on the notions of Kan extensions [13] and generalized representable functors [21] which prove to be very useful in this context. A generalized representable functor from a category $\underline{M}$ to a category $\underline{A}$ is a composite $A\otimes[M,-] : \underline{M} \longrightarrow \underline{S} \longrightarrow \underline{A}$, where $[M,-]$ is a hom-functor given by $M \in \underline{M}$ and $A\otimes$ the left adjoint of $[A,-] : \underline{A} \longrightarrow \underline{S}$. Recall that $A\otimes$ assigns to a set S the S-fold sum of A. Let $J : \underline{M} \longrightarrow \underline{C}$ be a functor and $\underline{A}$ be an abelian category with sums. The (right) Kan extension[2] $E_J : [\underline{M},\underline{A}] \longrightarrow [\underline{C},\underline{A}]$ is defined as the left adjoint of the "restriction" $R_J : [\underline{C},\underline{A}] \longrightarrow [\underline{M},\underline{A}], s \leadsto s\cdot J$. If $\underline{M}$ is not small, then $E_J(t)$ need not exist for every $t : \underline{M} \longrightarrow \underline{A}$ unless $\underline{C}$ has special properties (e.g. if there is a cotriple in $\underline{C}$). But if $t = A\otimes[M,-]$, then $E_J(t)$ exists and $E_J(t) = A\otimes[JM,-]$. The close relationship between Kan extensions and generalized representable functors is also illustrated by the fact that the functors $A\otimes[M,-]$ are acyclic objects for E_J; i.e. the higher left derived functors of E_J vanish on $A\otimes[M,-]$. If $\underline{C}$ is a category with a cotriple G and $\underline{M}$ is a subcategory of G-projectives,

---

[2] If J is full and faithful, then $E_J(t)$ is an extension of $t : \underline{M} \longrightarrow \underline{A}$. This explains the terminology.

then composite functors $t \cdot G : \underline{M} \longrightarrow \underline{M} \longrightarrow \underline{A}$ have properties analogous to the one of generalized representable functors.

The axioms of André and Barr-Beck for A_* and H_* essentially assert that they are the left derived functors of the Kan extension. This and the above mentioned properties of the functors $t \cdot G$ and $A \otimes [M,-]$ contain a lot of information about A_* and H_*.

A good deal of the material was first observed when André and Barr-Beck presented their non-abelian derived functors in seminars at the Forschungsinstitut during the winter of 1965-66 and summer of 1967.[3] In the meantime some of the results of this summary[4] were found independently by several authors. Among them are M. Bachmann (in a thesis under the supervision of B. Eckmann), E. Dubuc [9], U. Oberst [18], and D. Swan (unpublished). The paper of U. Oberst, which is to appear in the Math. Zeitschrift, led me to revise part of this note (or rather [23]) to include some of his results.

For the notation, terminology and the preliminaries

---

[3] Some of the material herein was first observed during the winter of 1967-68 after I had received an early version of [2]. The second half of [2] Ch. X was also developed during this period and illustrates the mutual influence of the material presented there and the corresponding material here and in [23].

[4] Details are to appear in another Lecture Notes volume [23].

about generalized representable functors and Kan extensions we refer the reader to [20] (1)-(12) which is contained in this Lecture Notes volume.

I am indebted to Jon Beck and Michael Barr for many stimulating discussions without which the paper would not have its present form.

Let $J : \underline{M} \longrightarrow \underline{C}$ be the inclusion of a small subcategory and let $\underline{A}$ be an abelian category with sums. For every functor $t \in [\underline{M},\underline{A}]$ and object $M \in \underline{M}$ there is a natural transformation $tM \otimes [M,-] \longrightarrow t$ which corresponds to id_{tM} under the Yoneda isomorphism $[tM \otimes [M,-], t] \cong [tM,tM]$. Its value at $X \in \underline{M}$ is a morphism $tM \otimes [M,X] = \underset{f}{\oplus} (tM)_f \longrightarrow tX$ which, restricted on a summand tM indexed by $f : M \longrightarrow X$, is $tf : tM \longrightarrow tX$. The family of identities $\{id_{tM}\}_{M \in \underline{M}}$ determines a natural transformation

$$(1) \qquad \phi(t) : \underset{M}{\oplus} (tM \otimes [M,-]) \longrightarrow t$$

which by the above is an objectwise split epimorphism. Since the representable functors $A \otimes [M,-]$ are projectives relative to the class $\mathcal{P}$ of short exact sequences in $[\underline{M},\underline{A}]$ which are objectwise split exact (cf. [20] (3), (4)), it follows that there are enough $\mathcal{P}$-projectives in $[\underline{M},\underline{A}]$. In [20] (12) we showed that $[\underline{M},\underline{A}]$ has also enough (absolute) projectives if $\underline{A}$ does. Thus we obtain from [20] (11) by standard homological algebra

(2) Theorem. Let $J : \underline{M} \longrightarrow \underline{C}$ be the inclusion of a small sub-category of $\underline{C}$ and $\underline{A}$ be an abelian category with sums. Then the Kan extension $E_J : [\underline{M},\underline{A}] \longrightarrow [\underline{C},\underline{A}]$ and its relative left derived functors $\mathcal{P}\text{-}L\, E_J$ exist. If $\underline{A}$ has either enough projectives or satisfies Grothendieck's axiom AB4),[5] then the absolute derived functors $L_* E_J$ also exist. In the latter case the functors $L_n E_J$ and $\mathcal{P}\text{-}L_n E_J$ coincide for $n \geq o$. In the following we denote $L_* E_J(-)$ by $A_*(\ ,-)$ and call it the André homology.

Proof. The only thing to prove is that the functors $\mathcal{P}\text{-}L_* E_J$ are the absolute derived functors of E_J if $\underline{A}$ is AB4). For every $t \in [\underline{M},\underline{A}]$ the epimorphism $\phi(t)$ in (1) gives rise to a relative projective resolution.

$$(3) \qquad P_*(t) : \ \ldots\ldots \bigoplus_{M \in \underline{M}} (t_1 M \otimes [M,-]) \longrightarrow \bigoplus_{M \in \underline{M}} (t_o M \otimes [M,-]) \xrightarrow{\phi(t)} t \longrightarrow 0$$

where $t_o = t$ and $t_1 = \ker \phi(t_o)$, etc.[6] Using [20] (9) and the property AB4) of $\underline{A}$, one can show that a short exact sequence of functors $0 \longrightarrow t' \longrightarrow t \longrightarrow t'' \longrightarrow 0$ in $[\underline{M},\underline{A}]$ gives rise to a short exact sequence $0 \longrightarrow E_J P_*(t') \longrightarrow E_J P_*(t) \longrightarrow E_J P_*(t'') \longrightarrow 0$ of chain complexes in $[\underline{C},\underline{A}]$. The long exact homology sequence associated with it makes $\mathcal{P}\text{-}L_* E_J$ into an absolute exact connected

---

[5] i.e. sums are exact in $\underline{A}$.

[6] $P_*(t)$ denotes the non-augmented complex, i.e. without t.

sequence of functors. Since $\mathfrak{P}\text{-}L_n E_J$ vanishes for $n > o$ on sums $\bigoplus_M (tM \otimes [M,-])$, it follows, by standard homological algebra, that $\mathfrak{P}\text{-}L_* E_J$ is left universal. In other words, the functors $\mathfrak{P}\text{-}L_n E_J$ are the (absolute) left derived functors $L_n E_J$ of the Kan extension $E_J : [\underline{M},\underline{A}] \longrightarrow [\underline{C},\underline{A}]$.

(4) A comparison with André's homology theory $H_*(\ ,-) : [\underline{M},\underline{A}] \longrightarrow [\underline{C},\underline{A}]$ in [1] p. 3-5, shows that, in view of his second axiom on page 15 [1], the functor $H_o(\ ,-)$ agrees with the Kan extension E_J on sums of representable functors.[7] Since both $H_o(\ ,-)$ and E_J are right exact, it follows from the exactness of (3) that they coincide. Since $H_n(\ ,-) : [\underline{M},\underline{A}] \longrightarrow [\underline{C},\underline{A}]$ vanishes for $n > o$ on sums of representable functors, it follows by standard homological algebra that the functors $H_*(\ ,-)$ are the left derived functors of E_J. Hence $H_*(\ ,-) \cong A_*(\ ,-) = L_* E_J(-)$ is valid. It may seem at first that this is "by chance" because André constructs H_* in an entirely different way. This however is not so. Recall that he associates with every functor $t : \underline{M} \longrightarrow \underline{A}$ a complex of functors $C_*(t) : \underline{C} \longrightarrow \underline{A}$ and defines $H_n(-,t)$ to be the n-th homology of $C_*(t)$ (cf. [1] p. 3).

---

[7] In the notation of André, $\underline{C}$ should be replaced by $\underline{N}$. Note that in view of [20] (9) a "foncteur élémentaire" of André is the Kan extension of a sum of representable functors $\underline{M} \longrightarrow \underline{A}$.

It is not difficult to show that the restriction of $C_n(t)$ on $\underline{M}$ is a sum of representable functors and that the Kan extension of $C_n(t)\cdot J$ is $C_n(t)$. Moreover, $C_*(t)\cdot J$ is an objectwise split exact resolution of t. Thus André's construction turns out to be the standard procedure in homological algebra to compute the left derived functors of E_J. Namely, choose an E_J-acyclic[8] resolution of t; apply E_J and take homology. The same is true for his computational device [1] prop. 1.5 because the restriction of the complex S_* on $\underline{M}$ is an E_J-acyclic resolution of $S\cdot J$ and $E_J(S_*\cdot J) = S_*$ is valid.[9] We now list some of the properties of $A_*(\ ,-)$ which, in part, generalize results of André [1]. They are consequences of (2), (3) and the nice behaviour of the Kan extension on representable functors.

(5) Theorem

(a) For every functor t the composite

$A_p(J-,t) : \underline{M} \longrightarrow \underline{C} \longrightarrow \underline{A}$ is zero for $p > o$.

---

[8] A functor is called E_J-acyclic if $L_nE_J : [\underline{M},\underline{A}] \longrightarrow [\underline{C},\underline{A}]$ vanishes on it for $n > o$.

[9] In [20] we show that this computational method is closely related with acyclic models.

(b) Assume that $\underline{A}$ is an AB5) category and let C be an object in $\underline{C}$ such that the comma category $(\underline{M},C)$ is directed.[10] Then $A_p(C,t)$ vanishes for $p > o$.

(c) Assume moreover that for every $M \in \underline{M}$ the hom-functor $[JM,-] : \underline{C} \longrightarrow \underline{S}$ preserves direct limits over directed index categories. Then $A_*(-,t) : \underline{C} \longrightarrow \underline{A}$ also preserves such direct limits. (In most examples this assumption is satisfied if the objects of $\underline{M}$ are finitely generated.)

If $\underline{M}$ has finite sums, it follows from (a) and (c) that $A_p(-,t) : \underline{C} \longrightarrow \underline{A}$ vanishes on arbitrary sums $\bigoplus_\nu M_\nu$, where $M_\nu \in \underline{M}$. As for applications, it is of great importance to establish this without assuming that $\underline{A}$ is AB5). We will sketch later how this can be done.

The properties (a) - (c) are straightforward consequences of (3), (2), footnote 15), [20] (9) and the fact that in an AB5) category direct limits over directed index categories are exact.

---

[10] A category $\underline{D}$ is called directed if for every pair D,D' of objects in $\underline{D}$ there is a $D'' \in \underline{D}$ together with morphisms $D \longrightarrow D''$, $D' \longrightarrow D''$ and if for a pair of morphisms $\gamma,\lambda: D_0 \longrightarrow D_1$ there is a morphism $\mu: D_1 \longrightarrow D_2$ such that $\mu\gamma = \mu\lambda$. Recall that objects of the category $(\underline{M},C)$ are morphisms $M \longrightarrow C$, where $M \in \underline{M}$.

(6) A change of models gives rise to a spectral sequence (cf. [1] prop. 8.1). For this, let $\underline{M}'$ be a small full sub-category of $\underline{C}$ containing $\underline{M}$. Denote by $J' : \underline{M}' \longrightarrow \underline{C}$ and $J'' : \underline{M} \longrightarrow \underline{M}'$ the inclusions and by A'_* and A''_* the associated André homologies. Since $J = J' \cdot J''$, it follows that $E_J : [\underline{M},\underline{A}] \longrightarrow [\underline{C},\underline{A}]$ is the composite of $E_{J''} : [\underline{M},\underline{A}] \longrightarrow [\underline{M}',\underline{A}]$ with $E_{J'} : [\underline{M}',\underline{A}] \longrightarrow [\underline{C},\underline{A}]$. Thus by standard homological algebra there is for every functor $t : \underline{M} \longrightarrow \underline{A}$ a spectral sequence

$$(7) \qquad A'_p(-,A''_q(-,t)) \Longrightarrow A_{p+q}(-,t)$$

provided $\underline{A}$ is AB4) or $\underline{A}$ has enough projectives (cf. [13] 2.4.1). One only has to verify that $E_{J''}$ takes E_J-acyclic objects into $E_{J'}$-acyclic objects. But this is obvious from [20] (9), because the Kan extension of a representable functor is again representable. The same holds for projective representable functors.[11]

The "Hochschild-Serre" spectral sequence of André [1] p. 33 can be obtained in the same way.

Likewise, a composed coefficient functor gives rise to a universal coefficient spectral sequence.

---

[11] The assumption that $\underline{M}'$ is small can be replaced by the following. The Kan extension $E_{J'} : [\underline{M}',\underline{A}] \longrightarrow [\underline{C},\underline{A}]$ and its left derived functors $L_* E_{J'}$ exist and $L_n E_{J'}$ vanishes on sums of representable functors for $n > o$.

(8) Theorem. Let $t : \underline{M} \longrightarrow \underline{A}$ and $F : \underline{A} \longrightarrow \underline{A}'$ be functors, where $\underline{A}$ and $\underline{A}'$ are either AB4) categories or have enough projectives. Assume that the left derived functors L_*F exist and that F has a right adjoint. Then there is a spectral sequence

$$L_pF \cdot A_q(-,t) \Longrightarrow A_{p+q}(-,F\cdot t)$$

provided the values of t are F-acyclic (i.e. $L_qF(tM) = 0$ for $q > o$).

(9) Corollary. If for every $p > o$ the functor $A_p(-,t) : \underline{C} \longrightarrow \underline{A}$ vanishes on an object $C \in \underline{C}$, then $A_*(C,F\cdot t) \cong L_*F(A_o(C,t))$. This gives rise to an infinite co-product formula, provided $A_o(-,t) : \underline{C} \longrightarrow \underline{A}$ is sum-preserving. For, let $C = \bigoplus_\nu C_\nu$ be an arbitrary sum with the property $A_p(C_\nu,t) = 0$ for $p > o$. Then the canonical map

(10) $$\bigoplus_\nu A_*(C_\nu,F\cdot t) \xrightarrow{\cong} A_*(\bigoplus_\nu C_\nu,F\cdot t)$$

is an isomorphism because $\bigoplus_\nu L_*F(A_o(C_\nu,F\cdot t)) \cong$ $\cong L_*F(\bigoplus_\nu A_o(C_\nu,F\cdot t)) \cong L_*F(A_o(\bigoplus_\nu C_\nu,F\cdot t))$ holds.[12]

---

[12] This applies to the categories of groups and semi-groups and yields infinite coproduct formulas for homology and co-homology of groups and semi-groups without conditions. For it can be shown that they coincide with A_* and A^* if $\underline{A}' = \underline{\text{Ab}}.\underline{\text{Gr}}.$, $\underline{A}$ = C-modules, t = Diff and F is tensoring with or homming into some C-module (cf. [2] Ch. 1 and Ch. X).

(11) Corollary. Assume that $\underline{A}$ is AB5) (but not $\underline{A}'$). Then every finite coproduct formula for $A_*(-,t) : \underline{C} \longrightarrow \underline{A}$ gives rise to an infinite coproduct formula for $A_*(-,F\cdot t) : \underline{C} \longrightarrow \underline{A}'$, provided the condition in (5c) is satisfied. In other words, $A_*(\bigoplus_\nu C_\nu, F\cdot t) \cong \bigoplus_\nu A_*(C_\nu, F\cdot t)$ holds if $A_*(\bigoplus_i C_{\nu_i}, t) \cong \bigoplus_i A_*(C_{\nu_i}, t)$ is valid for every finite sub-sum $\bigoplus_i C_{\nu_i}$ of $\bigoplus_\nu C_\nu$.[13]

Proof of (8) and (11). (Sketch) By $E_C : [\underline{C},\underline{A}] \longrightarrow \underline{A}$ and $E'_C : [\underline{C},\underline{A}'] \longrightarrow \underline{A}'$ we denote the evaluation of functors associated with $C \in \underline{C}$. The assumptions on F and t imply that the diagram

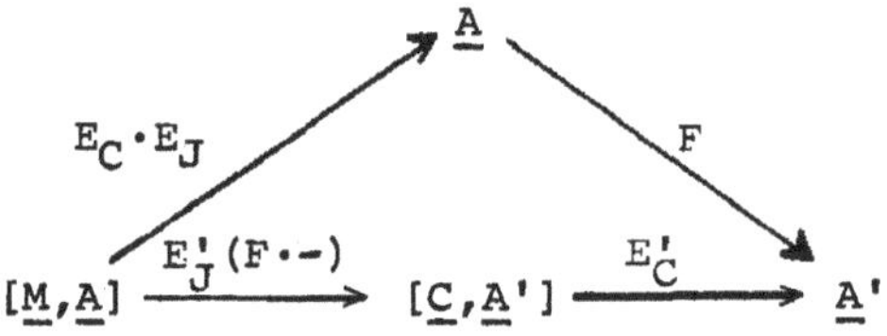

is commutative. The derived functors of $E'_C \cdot E'_J(f\cdot -) : [\underline{M},\underline{A}] \longrightarrow \underline{A}'$ can be identified with $t \rightsquigarrow A_*(C,F\cdot t)$. As above in (6) the spectral sequence arises from the decomposition of $E'_C \cdot E'_J(F\cdot -)$ into $E_C \cdot E_J$ and F. The infinite coproduct formula for $A_*(-,t) : \underline{C} \longrightarrow \underline{A}$ can be established by means of (5c).

---

[13] This applies to all finite coproduct theorems established in Barr-Beck [2] Ch. 7 and André [1] with $\underline{A}$, $\underline{A}'$, t and F as in footnote [12].

Thus it also holds for the E_2-term of the spectral sequence (8). One can show that the direct sum decomposition of the E_2-term is compatible with the differentials and the associated filtration of the spectral sequence. In this way, one obtains an infinite coproduct formula for

$A_*(-,F.t) : \underline{C} \longrightarrow \underline{A}'$.

(12) An abelian interpretation of André's non-abelian resolutions and neighborhoods is contained in a forthcoming paper of U. Oberst [18]. We include here a somewhat improved version of this interpretation and use it to solve a central problem which remained open in Barr-Beck [2] Ch. X. For this we briefly review the tensor product $\underline{\otimes}$ between functors which was investigated by D. Buchsbaum [5], J. Fisher [10], P. Freyd [11], D. Kan [13], U. Oberst [17] [18], C. Watts [24], N. Yoneda [25], and the author. The tensor product is a bifunctor

(13) $$\underline{\otimes} : [\underline{M}^{opp},\underline{Ab}.\underline{Gr}.] \times [\underline{M},\underline{A}] \longrightarrow \underline{A}$$

defined by the following universal property. For every $s \in [\underline{M}^{opp},\underline{Ab}.\underline{Gr}.]$, $t \in [\underline{M},\underline{A}]$ and $A \in \underline{A}$ there is an isomorphism

(14) $$[s \underline{\otimes} t,A] \cong [s,[t-,A]]$$

natural in s,t and A.

It can be constructed like the tensor product between Λ-modules, namely stepwise: 1) $\Lambda \otimes Y = Y$; 2) $(\bigoplus_\nu \Lambda_\nu) \otimes Y = \bigoplus_\nu Y_\nu$, where $\Lambda_\nu = \Lambda$ and $Y_\nu = Y$; 3) for an arbitrary module X choose a presentation $\bigoplus_\nu \Lambda_\nu \longrightarrow \bigoplus_\mu \Lambda_\mu \longrightarrow X \longrightarrow 0$ and define $X \otimes Y$ to be the co-kernel of the induced map $\bigoplus_\nu Y_\nu \longrightarrow \bigoplus_\mu Y_\mu$, where $Y_\nu = Y = Y_\mu$. The role of Λ is played by the family of contravariant representable functors $\mathbb{Z} \otimes [-,M] : \underline{M}^{opp} \longrightarrow \underline{Ab}.\underline{Gr}.$, where $M \in \underline{M}$ and $\mathbb{Z}$ denotes the integers.[14] Thus we define

$$(15) \qquad (\mathbb{Z} \otimes [-,M]) \underline{\otimes}\, t = tM$$

and continue as above. The universal property (13) follows from the Yoneda lemma [20] (2) in the following way:
$[\mathbb{Z} \otimes [-,M] \underline{\otimes}\, t,A] = [tM,A] \cong [\mathbb{Z},[tM,A]] \cong [\mathbb{Z} \otimes [-,M],[t-,A]]$.
One can show by means of (15) and the classical argument about balanced bifunctors that the derived functors of $\underline{\otimes}\, t : [\underline{M}^{opp},\underline{Ab}.\underline{Gr}.] \longrightarrow \underline{A}$ and $s\, \underline{\otimes} : [\underline{M},\underline{A}] \longrightarrow \underline{A}$ have the property $(L_* s\, \underline{\otimes})(t) \cong L_*(\underline{\otimes}\, t)(s)$, provided $\underline{A}$ is AB4) (resp. AB5)) and the values of s are free (resp. torsion free) abelian groups. Under these conditions the notion $\mathrm{Tor}_*(s,t)$ makes sense and has its usual properties, e.g. $\mathrm{Tor}_*(s,t)$ can be computed by projective or flat resolutions

---

[14] Note that the functors $\mathbb{Z} \otimes [-,M]$, $M \in \underline{M}$ are projective and form a generating family in $[\underline{M}^{opp},\underline{Ab}.\underline{Gr}.]$. This follows easily from the Yoneda lemma [20] (2) and (5).

in either variable. We remark without proof that every representable functor $A \otimes [M,-] : \underline{M} \longrightarrow \underline{A}$ is flat. It should be noted that for this and the following (until (22)) one cannot replace the condition AB4) by the assumption that $\underline{A}$ has enough projectives.

(16) U. Oberst [18] considers the class $\mathfrak{P}$ of short exact sequences in $[\underline{M}^{opp},\underline{Ab}.\underline{Gr}.]$ and $[\underline{M},\underline{A}]$ which are objectwise split exact. He shows that the derived functors of $s \underline{\otimes}$ and $\underline{\otimes}\, t$ relative to $\mathfrak{P}$ have the property $\mathfrak{P}\text{-}(L_* s\, \underline{\otimes})(t) \cong \mathfrak{P}\text{-}(L_*\, \underline{\otimes}\, t)(s)$ without any conditions on s and t. Thus the notion $\mathfrak{P}\text{-Tor}_*(s,t)$ makes sense. With every object $C \in \underline{C}$ there is associated a functor $\mathbf{Z} \otimes [J-,C] : \underline{M}^{opp} \longrightarrow \underline{Ab}.\underline{Gr}.$ the values of which are free abelian groups. (Recall that $J : \underline{M} \longrightarrow \underline{C}$ is the inclusion.) He establishes an isomorphism

(17) $$A_*(C,t) \cong \mathfrak{P}\text{-Tor}_*(\mathbf{Z} \otimes [J-,C],\ t)$$

for every functor $t : \underline{M} \longrightarrow \underline{A}$ and points out that a non-abelian resolution of C in the sense of André [1] p. 17 is a relative projective resolution of $\mathbf{Z} \otimes [J-,C]$. Thus André's result that $A_*(C,t)$ can be computed either by the complex $C_*(t) : \underline{C} \longrightarrow \underline{A}$ evaluated at C (cf. (4) and [1] p. 3) or a non-abelian resolution of C turns out to be a special case of the well known fact that $\mathfrak{P}\text{-Tor}_*(-,-)$ can be computed by a relative projective resolution of either variable.

U. Oberst also observes that a neighborhood ("voisinage") of C (cf. [1] p. 38) gives rise to a relative projective of $\mathbb{Z} \otimes [J-,C]$. Therefore, it is obvious that $A_*(C,t)$ can also be computed by means of neighborhoods.

(18) The notion of relative $\mathfrak{P}$-$Tor_*(-,-)$ is somewhat difficult to handle in practice. For instance, the spectral sequences (7) and (8) and the coproduct formulas (10) and (11) cannot be obtained with it because of the misbehaviour of the Kan extension on relative projective resolutions. Moreover, André's computational method [1] prop. 1.8 (cf. also (4)) cannot be explained by means of $\mathfrak{P}$-$Tor_*(\mathbb{Z} \otimes [J-,C],t)$ because the resolution of t in question need not be relative projective. Call a natural transformation $\phi : t \longrightarrow t'$ a $\mathfrak{P}$-natural transformation if for every $M \in \underline{M}$ the morphisms $tM \longrightarrow \mathrm{im}\phi(M) \longrightarrow 0$ and $0 \longrightarrow \mathrm{im}\phi(M) \longrightarrow t'M$ are split. The above difficulties arise because the Kan extensions $[\underline{M},\underline{A}] \longrightarrow [\underline{C},\underline{A}]$ and $[\underline{M}^{opp},\underline{Ab},\underline{Gr}.] \longrightarrow [\underline{C}^{opp},\underline{Ab}.\underline{Gr}.]$ do not preserve $\mathfrak{P}$-natural transformations. Our notion of absolute Tor_* does not have this disadvantage. The properties (17) etc. of the relative $\mathfrak{P}$-Tor can be established similarly for the absolute Tor_* using the techniques of U. Oberst [18]. We now sketch a different way to obtain these.

The fundamental relationship between $\underline{\otimes}$ and the Kan extension $E_J : [\underline{M},\underline{A}] \longrightarrow [\underline{C},\underline{A}]$ is given by the equation

(19) $$(\mathbb{Z} \otimes [J-,C]) \otimes t \cong E_J(t)(C)$$

where t and C are arbitrary objects of $[\underline{M},\underline{A}]$ and $\underline{C}$ respectively. To see this, let $[J-,C] = \varinjlim [-,M_\nu]$ be the canonical representation of $[J-,C] : \underline{M}^{opp} \longrightarrow \underline{S}$ as a direct limit of contravariant hom-functors. Note that the index category for this representation is isomorphic with the comma category $(\underline{M},C)$. Hence $\mathbb{Z} \otimes [J-,C] = \varinjlim \mathbb{Z} \otimes [-,M_\nu]$ and it follows from (15) and Kan's construction[15] that $(\mathbb{Z} \otimes [J-,C]) \underline{\otimes} t = \varinjlim tM_\nu = E_J(t)(C)$. Since $A_*(-,t)$ is the homology of the complex $E_J P_*(t)$, where $P_*(t)$ is the flat resolution (3) of t, it follows from (19) that

(20) $$A_*(C,t) \cong \mathrm{Tor}_*(\mathbb{Z} \otimes [J-,C],\ t)$$

Thus $A_*(C,t)$ can be computed either by projective resolutions of $\mathbb{Z} \otimes [J-,C]$ (e.g. non-abelian resolutions and neighborhoods[16]) or flat resolutions of t (e.g. $P_*(t)$ or André's resolutions $C_*(t)\cdot J$ and S_*, cf. (4)).

---

[15] Kan [13] 14 constructed $E_J : [\underline{M},\underline{A}] \longrightarrow [\underline{C},\underline{A}]$ objectwise. He showed that $E_J(t)(C)$ is the direct limit of $(\underline{M},C) \longrightarrow \underline{A}$, $(M \longrightarrow C) \rightsquigarrow tM$.

[16] Further examples are provided by the simplicial resolutions of Barr-Beck [2] Ch. 5, the projective simplicial resolutions of type (X,o) of Dold-Puppe [8] and the pseudo-simplicial resolutions of Tierney-Vogel [19]. A corollary of this is that the André (co)homology coincides with the theories developed by Barr-Beck [2], Dold-Puppe [8], and Tierney-Vogel [19] when $\underline{C}$ and $\underline{M}$ are defined appropriately. Note that there are more projective resolutions of $\mathbb{Z} \otimes [J-,C]$ than the ones described so far (for instance, the resolutions used in the proof of (21) below).

The above methods prove very useful in establishing the theorem below which is basic for many applications.

(21) Theorem. Let $\underline{M}$ be a full small sub-category of a category $\underline{C}$ which has sums. $\underline{M}$ need not have finite sums. However, if a sum $\bigoplus_i M_i \in \underline{C}$ is already in $\underline{M}$, it is assumed that every subsum of $\bigoplus_i M_i$ is also in $\underline{M}$. Moreover, assume that for every pair of objects $M \in \underline{M}$ and $\bigoplus_\nu M_\nu \in \underline{C}$ every morphism $M \longrightarrow \bigoplus_\nu M_\nu$ factors through a subsum belonging to $\underline{M}$, where $M_\nu \in \underline{M}$. Let $\underline{A}$ be an AB4) category and $t : \underline{M} \longrightarrow \underline{A}$ a sum-preserving functor.[17]

Then for $p > o$ the functor $A_p(-,t) : \underline{C} \longrightarrow \underline{A}$ vanishes on arbitrary sums $\bigoplus_\nu M_\nu$, where $M_\nu \in \underline{M}$.[18]

Proof. The idea is to construct a projective resolution $N_*(\bigoplus_\nu M_\nu, \overline{\underline{M}})$ of $\mathbf{Z} \otimes [J-, \bigoplus M_\nu] : \underline{M}^{opp} \longrightarrow \underline{Ab}.\underline{Gr}.$ which remains exact when tensored with $\underline{\otimes}\ t : [\underline{M}^{opp}, \underline{Ab}.\underline{Gr}.] \longrightarrow \underline{A}$.

With every pair of objects $C \in \underline{C}$ and $M \in \underline{M}$ and every full small sub-category $\overline{\underline{M}}$ of $\underline{M} \subset \underline{C}$,

---

[17] The meaning is that t has to preserve the sums which exist in $\underline{M}$.

[18] I am indebted to Michel André for an improvement of an earlier version of the theorem. He also found a different proof based on the methods he developed in [1], assuming that $\underline{M}$ has finite sums and that every morphism $M \longrightarrow \bigoplus M_\nu$ factors through a finite subsum.

André associated a s.s. set $\underline{\overline{M}}_*(M,C)$ together with an augmentation $\overline{M}_*(M,C) \longrightarrow [JM,C]$, (cf. [1] p. 38). It gives rise to an augmented complex of functors $\mathbb{Z} \otimes \underline{\overline{M}}_*(-,C) \longrightarrow \mathbb{Z} \otimes [J-,C]$ from $\underline{M}^{opp}$ to $\underline{Ab}.\underline{Gr}.$, where $\mathbb{Z} \otimes \underline{\overline{M}}_n(M,C)$ denotes the free abelian group on $\underline{\overline{M}}_n(M,C)$.

Let $C = \bigoplus_\nu M_\nu$, where $M_\nu \in \underline{M}$, and let $\underline{\overline{M}}$ be the full sub-category of $\underline{M}$, consisting of those subsums of $\bigoplus_\nu M_\nu$ which belong to $\underline{M}$. The resolution $N_*(\oplus M_\nu,\underline{\overline{M}})$ is defined to be a subcomplex of $\mathbb{Z} \otimes \underline{\overline{M}}_*(-,\bigoplus_\nu M_\nu)$ and consists in dimension n of a sum $\oplus (\mathbb{Z} \otimes [-,\overline{M}_n])$. More precisely, for every ascending chain of subsums $\overline{M}_n \Longrightarrow \overline{M}_{n-1} \Longrightarrow \ldots . \overline{M}_o \Longrightarrow \oplus M_\nu$ of $\oplus M_\nu$ there is a summand $\mathbb{Z} \otimes [-,\overline{M}_n]$, where $\overline{M}_i \in \underline{\overline{M}}$ for $o \leq i \leq n$ and the double arrows $\overline{M}_i \Longrightarrow \overline{M}_{i-1}$ and $\overline{M}_o \Longrightarrow \oplus M_\nu$ denote the canonical morphisms. The augmented complex $N_*(\oplus M_\nu,\underline{\overline{M}}) \longrightarrow \mathbb{Z} \otimes [J-,\oplus M_\nu]$ of functors $\underline{M}^{opp} \rightarrow \underline{Ab}.\underline{Gr}.$ is objectwise split exact. The contraction $s_n(M) : N_n(\oplus M_\nu,\underline{\overline{M}})(M) \longrightarrow N_{n+1}(\oplus M_\nu,\underline{\overline{M}})(M)$ is defined as follows. For a sum $\bigoplus_\nu M_\nu$ and a subsum $\bigoplus_i M_{\nu_i}$ denote by $\bigoplus_i M_{\nu_i} \Longrightarrow \bigoplus_\nu M_\nu$ the canonical morphism, $M_\nu \in \underline{M}$. Let $f : M \longrightarrow \bigoplus_\nu M_\nu$ be a morphism, $M_\nu \in \underline{M}$. Then there is a unique decomposition $M \longrightarrow M^f \Longrightarrow \oplus M_\nu$ of f such that $M^f \Longrightarrow \oplus M_\nu$ is the smallest subsum of $\bigoplus_\nu M_\nu$ which contains the "image" of f. Obviously M^f is an object of $\underline{\overline{M}}$. The morphism $s_{-1}(M) : \mathbb{Z} \otimes [JM, \bigoplus_\nu M_\nu] \longrightarrow \oplus (\mathbb{Z} \otimes [M,\overline{M}_o])$ assigns to a base element $f : M \longrightarrow \oplus M_\nu$ its factorization $M \longrightarrow M^f$ in the component $\mathbb{Z} \otimes [M,M^f]$ indexed by $M^f \longrightarrow \oplus M_\nu$.

Likewise, $s_n(M) : N_n(\bigoplus_\nu M_\nu,\underline{\overline{M}})(M) \longrightarrow N_{n+1}(\bigoplus_\nu M_\nu,\underline{\overline{M}})(M)$ assigns to a base element $f : M \longrightarrow \overline{M}_n$ of the summand $\mathbb{Z} \otimes [M,\overline{M}_n]$ indexed by $\overline{M}_n \Longrightarrow \dots \overline{M}_o \Longrightarrow \bigoplus_\nu M_\nu$ its factorization $M \longrightarrow M^f$ in the component $\mathbb{Z} \otimes [M,M^f]$ indexed by $M^f \Longrightarrow \overline{M}_n \Longrightarrow \dots \overline{M}_o \Longrightarrow \bigoplus_\nu M_\nu$. Since M^f and $\overline{M}_i$ are subsums of $\bigoplus_\nu M_\nu$, where $o \leq i \leq n$, it is not difficult to check that $s_*(M)$ is a contraction of $N_*(\bigoplus_\nu M_\nu,\underline{\overline{M}})(M) \longrightarrow \mathbb{Z} \otimes [JM, \bigoplus_\nu M_\nu]$. (However $s_n(M)$ is not natural in M .) Hence $N_*(\bigoplus_\nu M_\nu,\underline{\overline{M}})$ is a projective resolution of $\mathbb{Z} \otimes [J-, \bigoplus M_\nu]$ and the p-th homology of the complex $N_*(\bigoplus_\nu M_\nu,\underline{M}) \underline{\otimes}\, t$ is $A_p(\bigoplus_\nu M_\nu,t)$. By (15) the latter consists of a sum

$$\bigoplus_{\overline{M}_n \Longrightarrow \dots \overline{M}_o \Longrightarrow \bigoplus_\nu M_\nu} t\overline{M}_n$$

in dimension n. To prove that this complex has trivial homology except in dimension zero, I originally assumed that a certain s.s. set satisfies the Kan condition (which is present in all examples I know). M. André then pointed out to me that this condition is redundant because the above subcomplex has a contraction h_* which is defined as follows. Recall that the object $\overline{M}_n$ of an ascending chain $\overline{M}_n \Longrightarrow \dots \overline{M}_o \Longrightarrow \bigoplus M_\nu$ is a subsum $\bigoplus_i M_{\nu_i}$ of $\bigoplus_\nu M_\nu$. Denote by $M_{\nu_i} \Longrightarrow \overline{M}_n$ the canonical morphism. The morphism $h_n : \bigoplus t\overline{M}_n \longrightarrow \bigoplus t\overline{M}_{n+1}$ maps the summand tM_{ν_i} of the component $t\overline{M}_n = \bigoplus_i tM_{\nu_i}$ indexed by $\overline{M}_n \Longrightarrow \overline{M}_{n-1} \Longrightarrow \dots \overline{M}_o \Longrightarrow \bigoplus M_\nu$ identically on the component tM_{ν_i} of $\oplus\, t\overline{M}_{n+1}$ indexed by $M_{\nu_i} \Longrightarrow \overline{M}_n \Longrightarrow \overline{M}_{n-1} \Longrightarrow \dots \overline{M}_o \Longrightarrow \underset{\nu}{\oplus} M_\nu$. It is routine to check that h_* is a contraction. This shows that $A_p(\underset{\nu}{\oplus} M_\nu, t) = 0$ for $p > o$.

Q.E.D.

(22) Corollary. Let $\underline{M}'$ be the full sub-category of $\underline{C}$ consisting of sums of objects in $\underline{M}$. Assume that the Kan extension $E_{J'} : [\underline{M}', \underline{A}] \longrightarrow [\underline{C}, \underline{A}]$ and its left derived functors $L_* E_{J'} = A'_*$ exist and that $L_n E_{J'}$ vanishes on representable functors for $n > o$. Then by (21) the spectral sequence (7) collapses and one obtains an isomorphism

$$A_*(-,t) \xrightarrow{\cong} A'_*(-, A''_o(-,t))$$

where $t : \underline{M} \longrightarrow \underline{A}$ is a functor as in (21) and $A''_o(-,t) : \underline{M}' \longrightarrow \underline{A}$ is its Kan extension on $\underline{M}'$.[19]

The value of (22) lies in the fact that $A' : \underline{C} \longrightarrow \underline{A}$ can be identified with the homology associated with a certain cotriple in $\underline{C}$ (the model induced cotriple (cf. (33) (34)). In this way every André homology can be realized as a cotriple homology and all information about the latter carries over to the former and vice versa.

---

[19] If $\underline{A}$ is AB5), then the theorem is also true if $\underline{M}'$ is an arbitrary full sub-category of $\underline{C}$, such that for every $M' \in \underline{M}'$ the associated comma category $(\underline{M}, M')$ is directed. This follows easily from (5b) and (7).

It has become apparent in several places that the smallness of $\underline{M}$ is an unpleasant restriction which should be removed. André did this by requiring that every $C \in \underline{C}$ has a neighborhood in $\underline{M}$. Another way of expressing the same condition is to assume that every functor $\mathbb{Z} \otimes [J-,C] : \underline{M}^{opp} \longrightarrow \underline{Ab}.\underline{Gr}.$ admits a projective resolution. It is then clear from the above that there is an exact connected sequence of functors $A_*(\ ,-) : [\underline{M},\underline{A}] \longrightarrow [\underline{C},\underline{A}]$ with the properties $A_o(\ ,-) = E_J$ and $A_n(-,A \otimes [-,M]) = 0$ for $n > o$. Since $\underline{M}$ is not small, not every functor $t : \underline{M} \longrightarrow \underline{A}$ need be a quotient of a sum of representable functors and one cannot automatically conclude that $L_*E_J = A_*$. In many examples this is however the case, e.g. if $\underline{M}$ consists of the G-projectives of a cotriple G in $\underline{C}$.

(23) So far we have only dealt with André homology with respect to the inclusion $J : \underline{M} \longrightarrow \underline{C}$ and not with the homology associated with a cotriple G in C (for the definition of a cotriple we refer to Barr-Beck [2] intro.). One reason for this is that the corresponding model category $\underline{M}$ is not small. Another is that the presence of a cotriple is a more special situation in which theorems often hold under weaker conditions and proofs are easier. The additional information is due to the simple behaviour of the Kan extension on functors of the form $\underline{M} \xrightarrow{G} \underline{M} \xrightarrow{t} \underline{A}$, where G is the restriction of the cotriple on $\underline{M}$. We now outline how our approach works for cotriple homology.

(24) Let G be a cotriple in $\underline{C}$ and denote by $\underline{M}$ any full sub-category of $\underline{C}$, the objects of which are G-projectives and include every GC, where $C \in \underline{C}$. (Recall that an object $X \in \underline{C}$ is called G-projective if $\varepsilon(X) : GX \longrightarrow X$ admits a section, where $\varepsilon : G \longrightarrow id_{\underline{C}}$ is the co-unit of the cotriple. The objects GC are called free.) With every functor $t : \underline{C} \longrightarrow \underline{A}$ Barr-Beck [2] associate the cotriple derived functors $H_*(-,t)_G : \underline{C} \longrightarrow \underline{A}$, also called cotriple homology with coefficient functor t. Their construction of $H_*(-,t)_G$ only involves the values of t on the free objects of $\underline{C}$. Thus $H_*(-,t)_G$ is also well defined when t is only defined on $\underline{M}$.

(25) Theorem. The Kan extension $E_J : [\underline{M},\underline{A}] \longrightarrow [\underline{C},\underline{A}]$ exists. It assigns to a functor $t : \underline{M} \longrightarrow \underline{A}$ the zeroth cotriple derived functor $H_0(-,t)_G : \underline{C} \longrightarrow \underline{A}$. In particular $E_J(t \cdot G) = t \cdot G$ is valid. (Note that $\underline{A}$ need not be AB3) or AB4) for this.)

Proof. According to the definition of a Kan extension, we have to show that for every $S : \underline{C} \longrightarrow \underline{A}$ the restriction map $[H_0(-,t)_G, S] \longrightarrow [t, S \cdot J]$ is a bijection. We limit ourselves to giving a map in the opposite direction and leave it to the reader to check that they are inverse to each other. A natural transformation $\phi : t \longrightarrow S \cdot J$ gives rise to a diagram

$$\begin{array}{ccccc}
tG^2C & \overset{t\varepsilon(GC)}{\underset{tG(\varepsilon C)}{\rightrightarrows}} & tGC & \longrightarrow & H_o(C,t)_G \\
\downarrow{\scriptstyle \phi(G^2C)} & & \downarrow{\scriptstyle \phi(GC)} & & \\
SG^2C & \overset{S\varepsilon(GC)}{\underset{SG\varepsilon(C)}{\rightrightarrows}} & SGC & \xrightarrow{S\varepsilon(C)} & SC
\end{array}$$

for every $C \in \underline{C}$. The top row is by construction of $H_o(C,t)_G$ a coequalizer. Thus there is a unique morphism $H_o(C,t)_G \longrightarrow SC$ which makes the diagram commutative. In this way one obtains a natural transformation $H_o(-,t)_G \longrightarrow S$.

The properties established in [2] Ch. I imply that $H_*(\ ,-)_G : [\underline{M},\underline{A}] \longrightarrow [\underline{C},\underline{A}]$ is an exact connected sequence of functors. Since $H_n(-,t\cdot G) = 0$ for every functor $t : \underline{M} \longrightarrow \underline{A}$ and since the canonical natural transformation $tG \longrightarrow t$ is an (objectwise split) epimorphism, we obtain by standard homological algebra the following

(26) Theorem. The left derived functors of the Kan extension $E_J : [\underline{M},\underline{A}] \longrightarrow [\underline{C},\underline{A}]$ exist and $L_*E_J(-) \cong H_*(\ ,-)_G$ is valid. Moreover $\mathfrak{P}\text{-}L_*E_J(-) \cong H_*(\ ,-)_G$ holds, where $\mathfrak{P}$ denotes the class of short exact sequences in $[\underline{M},\underline{A}]$ which are objectwise split exact. We remark without proof that for $n > o$, $H_n(\ ,-)_G$ vanishes on sums of representable functors.

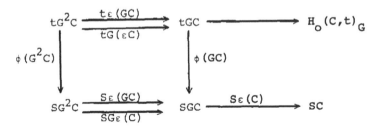

(27) Corollary. The cotriple homology depends only on the G-projectives.[20] In particular, two cotriples G and G' in $\underline{C}$ give rise to the same homology if their projectives coincide. One can show that the converse is also true.

(28) It is obvious from the above that the axioms of Barr-Beck [2] Ch. III for $H_*(-,-)_G$ are the usual acyclicity criterion for establishing the universal property of an exact connected sequence of functors. As in (4) the constructions of $H_*(-,t)_G$ in [2] by means of the s.s. resolution $tG_* : \underline{C} \longrightarrow \underline{A}$ is actually the standard procedure in homological algebra. This is because the restriction of tG_* on $\underline{M}$ is an E_J-acyclic resolution of t and because $E_J(tG_* \cdot J) = tG_*$ holds (for G_* see [2] Chap. I).

(29) It also follows from (26) that cotriple homology $H_*(-,t)_G$ and André homology $A_*(-,t)$ coincide, provided the models for the latter are $\underline{M}$. One might be tempted to deduce this from the first half of (26), but apparently it can only be obtained from the second half. The reason is a set theoretical difficulty. For details we refer to [23].

---

[20] In many cases the cotriple homology depends only on the finitely generated G-projectives. An object $X \in \underline{C}$ is called finitely generated if the hom-functor $[X,-] : \underline{C} \longrightarrow \underline{S}$ preserves filtered unions of subobjects.

From this it is obvious that the properties previously established for the André homology carry over to the cotriple homology. We list below some useful modifications which result from direct proofs of these properties.

(30) The assumption in (5c), which is seldom present in examples when $\underline{M}$ is not small, can be replaced by the following: The cotriple $G : \underline{C} \longrightarrow \underline{M}$ and the functor $t : \underline{M} \longrightarrow \underline{A}$ preserve directed direct limits.

In (8)-(11) the functor F need not have a right adjoint. It suffices instead that F is right exact.

From (26), (7), and footnote [11] we obtain for a small sub-category $\overline{\underline{M}}$ of $\underline{M}$ a spectral sequence

(31) $$H_p(-,\overline{A}_q(-,t))_G \Longrightarrow A_{p+q}(-,t)$$

where $\underline{M}$ is as in (24), and $\overline{A}_*$ and A_* denote the left derived functors of the Kan extensions $[\overline{\underline{M}},\underline{A}] \longrightarrow [\underline{M},\underline{A}]$ and $[\underline{M},\underline{A}] \longrightarrow [\underline{C},\underline{A}]$ respectively.

The tensor product
$\underline{\otimes} : [\underline{M}^{opp},\underline{Ab}.\underline{Gr}.] \times [\underline{M},\underline{A}] \longrightarrow \underline{A}$, $s x t \rightsquigarrow s \underline{\otimes} t$
is defined as in (14) but may not exist for every $s \in [\underline{M}^{opp},\underline{Ab}.\underline{Gr}.]$. However (19) and (32)
$$\mathrm{Tor}_*(\mathbb{Z} \otimes [J-,C],\ t) \cong H_*(C,t)_G \cong \mathfrak{p}\text{-}\mathrm{Tor}_*(\mathbb{Z} \otimes [J-,C],\ t)$$
hold.

The first isomorphism shows that $H_*(C,t)_G$ can be computed by either a projective resolution of $\mathbb{Z} \otimes [J-,C]$ or an E_J-acyclic resolution of t. The former is a generalization of the main result in [2] 5.1, because a G-resolution is a projective resolution of $\mathbb{Z} \otimes [J-,C]$; the latter generalizes the acyclic model argument in [4] (cf. also [20] (25)).

(33) With every small sub-category $\underline{M}$ of a category $\underline{C}$ there is associated a cotriple G, called the model induced cotriple (cf. [2] Ch. X). Recall that its functor part $G : \underline{C} \longrightarrow \underline{C}$ assigns to an object $C \in \underline{C}$ the sum $\bigoplus_f df$, indexed by morphisms $f : df \longrightarrow C$, the domain df of which belongs to $\underline{M}$. The co-unit $\varepsilon(C) : \oplus\, df \longrightarrow C$ restricted on a summand df is $f : df \longrightarrow C$. The theorems (26) and (22) enable us to compare the André homology $A_*(-,-)$ of the inclusion $\underline{M} \longrightarrow \underline{C}$ with the homology of the model induced cotriple G in $\underline{C}$. Since every sum $\bigoplus_\nu M_\nu$ of objects $M_\nu \in \underline{M}$ is G-projective, we obtain the following:

(34) <u>Theorem</u>. Assume that the inclusion $\underline{M} \longrightarrow \underline{C}$ satisfies the conditions in (21). Then for every sum-preserving functor $t : \underline{M} \longrightarrow \underline{A}$ the canonical map

$$A_*(-,t) \xrightarrow{\cong} H_*(-,A_o(-,t))_G$$

is an isomorphism, provided $\underline{A}$ is an AB4) category.

If t is contravariant and takes sums into produces, we obtain likewise for cohomology

$$A^*(-,t) \xleftarrow{\cong} H^*(-,A^o(-,t))_G$$

provided $\underline{A}$ is an AB4)* category.

The theorem asserts that André (co)homology can be realized under rather weak conditions as (co)homology of a cotriple, even of a model induced cotriple. In this way, the considerations in Barr-Beck [2] 7.1, 7.2 (coproduct formulas), [2] 8.1 (homology sequence of a map) and [2] 9.1, 9.2 (Mayer Vietoris) also apply to André (co)homology. Moreover, the fact that cotriple cohomology tends to classify extensions (cf. [3]) carries over to André cohomology. We illustrate the use of this realization with some examples.

(35) Examples

a) Let $\underline{C}$ be a category of algebras with rank $(\underline{C}) = \alpha$ in the sense of Linton [15]. Recall that if $\alpha = \aleph_0$, then $\underline{C}$ is a category of universal algebras in the classical sense (cf. Lawvere [14]). By means of (34) and (27) one can show that (co)homology of the free cotriple in $\underline{C}$ coincides with the André (co)homology associated with inclusion $\underline{M} \longrightarrow \underline{C}$, where the objects of $\underline{M}$ are free algebras on less than α generators.

b) Let $\underline{C}$ = $\underline{Ab}.\underline{Gr}.$ and let $\underline{M}$ be the sub-category of finitely generated abelian groups. Using (34), one can show that the first André cohomology group $A^1(C,[-,Y])$ is isomorphic to the group of pure extensions of $C \in \underline{Ab}.\underline{Gr}.$ with kernel $Y \in \underline{Ab}.\underline{Gr}.$ in the sense of Harrison. The same holds if $\underline{C}$ is a category of Λ-modules, where Λ is a ring with unit.

c) Let $\underline{C}$ = Λ-algebras and let $\underline{M}$ be the sub-category of finitely generated tensor algebras. Let C be a Λ-algebra and Y be a Λ-bimodule. Then $A^1(C,[-,Y])$ classifies singular extensions $E \longrightarrow C$ with kernel Y such that the underlying Λ-module extension is pure in the sense of Harrison (cf. b)).[21]

The cases (b) and (c) set the tone for a long list of similar examples, which indicate that Harrison's theory of pure group extensions can be considerably generalized.

---

[21] I am indebted to M. Barr for correcting an error I had made in this example.

BIBLIOGRAPHY

[1] André, M., *Méthode simpliciale en algèbre homologique et algèbre commutative*, (Lecture notes in Mathematics, #32), Springer, 1967.

[2] Barr, M. and J. Beck, "Homology and standard constructions", (to appear in Lecture Notes in Mathematics).

[3] Beck, J., "Triples, algebras and cohomology", *Dissertation*, Columbia, (1964-67).

[4] Barr, M. and Beck, J., "Acyclic models and triples", in; *Conference on Categorical Algebra*, pp. 336-343, Springer, (1966).

[5] Buchsbaum, D., "Homology and universal functors", in; (Lecture Notes in Mathematics, #61), Springer, 1968.

[6] Dold, A., "Zur Homotopietheorie der Kettenkomplexe", *Math. Annalen*, 140; 278-298, (1960).

[7] Dold, A., S. MacLane, U. Oberst, "Projective classes and acyclic models", in; A. Dold, Heidelberg and Eckmann (eds.), *Reports of the Midwest Category Seminar*, (Lecture Notes in Mathematics, #47); 78-91, Springer, 1967.

[8] Dold, A. and D. Puppe, "Homologie nicht-additiver Funktoren", *Ann. Inst. Fourier*, 11; 291- (1961).

[9] Dubuc, E., "Adjoint triangles", in; (Lecture notes in Mathematics, #61), Springer, 1968.

[10] Fisher, J., "The tensor product of functors, satellites and derived functors", *J. of Algebra*, to appear.

[11] Freyd, P., *Abelian Categories*, Harper and Row: New York, 1964.

[12] Grothendieck, A., "Sur quelques points d'algèbre homologique", *Tôhoku, Math. J.*, 9; 119-221, (1957).

[13] Kan, D., "Adjoint Functors", *Trans. Amer. Math. Soc.*, 87; 295-329, (1958).

[14] Lawvere, F., "Functorial semantics and algebraic theories", *Proc. Nat. Acad. Sci.*, 50; 869-872, (1963).

[15] Linton, F., "Some aspects of equational categories", in; *Conference on Categorical Algebra*, pp. 84-94, Springer, (1966).

[16] MacLane, S., *Homology*, Springer, 1963.

[17] Oberst, U., "Basiserweiterung in der Homologie kleiner Kategorien", *Math. Zeitschrift*, 100; 36-58, (1967).

[18] Oberst, U., "Homology of categories and exactness of direct limits", to appear.

[19] Tierney, M., and W. Vogel, "Simplicial resolutions and derived functors", *Mim. Notes*, E.T.H., (1968).

[20] Ulmer, F., "Acyclic models and Kan extensions", this volume.

[21] ______. "Representable functors with values in arbitrary categories", *J. of Algebra*, 8; 96-129, (1968).

[22] ______. "Properties of Kan extensions", *Mim. Notes*, E.T.H., (1966).

[23] ______. "On André and cotriple (co)homology and their relationship to classical homological algebra", (to appear in Lecture Notes in Mathematics).

[24] Watts, C. E., "A homology theory for small categories", in; *Conference on Categorical Algebra*, pp. 331-335, Springer, (1966).

[25] Yoneda, N., "On Ext and exact sequences", *J. Fac. Sci. Tokyo, Sec. I*, 8; 507-576, (1961).

Offsetdruck: Julius Beltz, Weinheim/Bergstr.

Lecture Notes in Mathematics

Bisher erschienen/Already published

Vol. 1: J. Wermer, Seminar über Funktionen-Algebren. IV, 30 Seiten. 1964. DM 3,80 / 0.95

Vol. 2: A. Borel, Cohomologie des espaces localement compacts d'après J. Leray. IV, 93 pages. 1964. DM 9,– / $ 2.25

Vol. 3: J. F. Adams, Stable Homotopy Theory. 2nd. revised edition. IV, 78 pages. 1966. DM 7,80 / $ 1.95

Vol. 4: M. Arkowitz and C. R. Curjel, Groups of Homotopy Classes. 2nd. revised edition. IV, 36 pages. 1967. DM 4,80 / $ 1.20

Vol. 5: J.-P. Serre, Cohomologie Galoisienne. Troisième édition. VIII, 214 pages. 1965. DM 18,– / $ 4.50

Vol. 6: H. Hermes, Eine Termlogik mit Auswahloperator. IV, 42 Seiten. 1965. DM 5,80 / $ 1.45

Vol. 7: Ph. Tondeur, Introduction to Lie Groups and Transformation Groups. VIII, 176 pages. 1965. DM 13,50 / $ 3.40

Vol. 8: G. Fichera, Linear Elliptic Differential Systems and Eigenvalue Problems. IV, 176 pages. 1965. DM 13,50 / $ 3.40

Vol. 9: P. L. Ivănescu, Pseudo-Boolean Programming and Applications. IV, 50 pages. 1965. DM 4,80 / $ 1.20

Vol. 10: H. Lüneburg, Die Suzukigruppen und ihre Geometrien. VI, 111 Seiten. 1965. DM 8,– / $ 2.00

Vol. 11: J.-P. Serre, Algèbre Locale. Multiplicités. Rédigé par P. Gabriel. Seconde édition. VIII, 192 pages. 1965. DM 12,– / $ 3.00

Vol. 12: A. Dold, Halbexakte Homotopiefunktoren. II, 157 Seiten. 1966. DM 12,– / $ 3.00

Vol. 13: E. Thomas, Seminar on Fiber Spaces. IV, 45 pages. 1966. DM 4,80 / $ 1.20

Vol. 14: H. Werner, Vorlesung über Approximationstheorie. IV, 184 Seiten und 12 Seiten Anhang. 1966. DM 14,– / $ 3.50

Vol. 15: F. Oort, Commutative Group Schemes. VI, 133 pages. 1966. DM 9,80 / $ 2.45

Vol. 16: J. Pfanzagl and W. Pierlo, Compact Systems of Sets. IV, 48 pages. 1966. DM 5,80 / $ 1.45

Vol. 17: C. Müller, Spherical Harmonics. IV, 46 pages. 1966. DM 5,– / $ 1.25

Vol 18: H.-B. Brinkmann und D. Puppe, Kategorien und Funktoren. XII, 107 Seiten, 1966. DM 8,– / $ 2.00

Vol. 19: G. Stolzenberg, Volumes, Limits and Extensions of Analytic Varieties. IV, 45 pages. 1966. DM 5,40 / $ 1.35

Vol. 20: R. Hartshorne, Residues and Duality. VIII, 423 pages. 1966. DM 20,– / $ 5.00

Vol. 21: Seminar on Complex Multiplication. By A. Borel, S. Chowla, C. S. Herz, K. Iwasawa, J.-P. Serre. IV, 102 pages. 1966. DM 8,– / $ 2.00

Vol. 22: H. Bauer, Harmonische Räume und ihre Potentialtheorie. IV, 175 Seiten. 1966. DM 14,– / $ 3.50

Vol. 23: P. L. Ivănescu and S. Rudeanu, Pseudo-Boolean Methods for Bivalent Programming. 120 pages. 1966. DM 10,– / $ 2.50

Vol. 24: J. Lambek, Completions of Categories. IV, 69 pages. 1966. DM 6,80 / $ 1.70

Vol. 25: R. Narasimhan, Introduction to the Theory of Analytic Spaces. IV, 143 pages. 1966. DM 10,– / $ 2.50

Vol. 26: P.-A. Meyer, Processus de Markov. IV, 190 pages. 1967. DM 15,– / $ 3.75

Vol. 27: H. P. Künzi und S. T. Tan, Lineare Optimierung großer Systeme. VI, 121 Seiten. 1966. DM 12,– / $ 3.00

Vol. 28: P. E. Conner and E. E. Floyd, The Relation of Cobordism to K-Theories. VIII, 112 pages. 1966. DM 9,80 / $ 2.45

Vol. 29: K. Chandrasekharan, Einführung in die Analytische Zahlentheorie. VI, 199 Seiten. 1966. DM 16,80 / $ 4.20

Vol. 30: A. Frölicher and W. Bucher, Calculus in Vector Spaces without Norm. X, 146 pages. 1966. DM 12,– / $ 3.00

Vol. 31: Symposium on Probability Methods in Analysis. Chairman. D. A. Kappos. IV, 329 pages. 1967. DM 20,– / $ 5.00

Vol. 32: M. André, Méthode Simpliciale en Algèbre Homologique et Algèbre Commutative. IV, 122 pages. 1967. DM 12,– / $ 3.00

Vol. 33: G. I. Targonski, Seminar on Functional Operators and Equations. IV, 110 pages. 1967. DM 10,– / $ 2.50

Vol. 34: G. E. Bredon, Equivariant Cohomology Theories. VI, 64 pages. 1967. DM 6,80 / $ 1.70

Vol. 35: N. P. Bhatia and G. P. Szegö, Dynamical Systems. Stability Theory and Applications. VI, 416 pages. 1967. DM 24,– / $ 6.00

Vol. 36: A. Borel, Topics in the Homology Theory of Fibre Bundles. VI, 95 pages. 1967. DM 9,– / $ 2.25

Vol. 37: R. B. Jensen, Modelle der Mengenlehre. X, 176 Seiten. 1967. DM 14,– / $ 3.50

Vol. 38: R. Berger, R. Kiehl, E. Kunz und H.-J. Nastold, Differentialrechnung in der analytischen Geometrie IV, 134 Seiten. 1967. DM 12,– / $ 3.00

Vol. 39: Séminaire de Probabilités I. II, 189 pages. 1967. DM 14,– / $ 3.50

Vol. 40: J. Tits, Tabellen zu den einfachen Lie Gruppen und ihren Darstellungen. VI, 53 Seiten. 1967. DM 6.80 / $ 1.70

Vol. 41: A. Grothendieck, Local Cohomology. VI, 106 pages. 1967. DM 10.– / $ 2.50

Vol. 42: J. F. Berglund and K. H. Hofmann, Compact Semitopological Semigroups and Weakly Almost Periodic Functions. VI, 160 pages. 1967. DM 12,– / $ 3.00

Vol. 43: D. G. Quillen, Homotopical Algebra VI, 157 pages. 1967. DM 14,– / $ 3.50

Vol. 44: K. Urbanik, Lectures on Prediction Theory IV, 50 pages. 1967. DM 5,80 / $ 1.45

Vol. 45: A. Wilansky, Topics in Functional Analysis VI, 102 pages. 1967. DM 9,60 / $ 2.40

Vol. 46: P. E. Conner, Seminar on Periodic Maps IV, 116 pages. 1967. DM 10,60 / $ 2.65

Vol. 47: Reports of the Midwest Category Seminar I. IV, 181 pages. 1967. DM 14,80 / $ 3.70

Vol. 48: G. de Rham, S. Maumary et M. A. Kervaire, Torsion et Type Simple d'Homotopie. IV, 101 pages. 1967. DM 9,60 / $ 2.40

Vol. 49: C. Faith, Lectures on Injective Modules and Quotient Rings. XVI, 140 pages. 1967. DM 12,80 / $ 3.20

Vol. 50: L. Zalcman, Analytic Capacity and Rational Approximation, VI, 155 pages. 1968. DM 13.20 / $ 3.40

Vol. 51: Séminaire de Probabilités II. IV, 199 pages. 1968. DM 14,– / $ 3.50

Vol. 52: D. J. Simms, Lie Groups and Quantum Mechanics. IV, 90 pages. 1968. DM 8,– / $ 2.00

Vol. 53: J. Cerf, Sur les difféomorphismes de la sphère de dimension trois ($\Gamma_4 = 0$). XII, 133 pages. 1968. DM 12,– / $ 3.00

Vol. 54: G. Shimura, Automorphic Functions and Number Theory VI, 69 pages. 1968. DM 8,– / $ 2.00

Vol. 55: D. Gromoll, W. Klingenberg und W. Meyer Riemannsche Geometrie im Großen VI, 287 Seiten. 1968. DM 20,– / $ 5.00

Bitte wenden / Continued

Vol. 56: K. Floret und J. Wloka, Einführung in die Theorie der lokalkonvexen Räume VIII, 194 Seiten. 1968. DM 16,– / $ 4.00

Vol. 57: F. Hirzebruch und K. H. Mayer, O(n)-Mannigfaltigkeiten, exotische Sphären und Singularitäten. IV, 132 Seiten. 1968. DM 10,80 / $ 2.70

Vol. 58: Kuramochi Boundaries of Riemann Surfaces. IV, 102 pages. 1968. DM 9,60 / $ 2.40

Vol. 59: K. Jänich, Differenzierbare G-Mannigfaltigkeiten. VI, 89 Seiten. 1968. DM 8,– / $ 2.00

Vol. 60: Seminar on Differential Equations and Dynamical Systems. Edited by G. S. Jones VI, 106 pages. 1968. DM 9,60 / $ 2.40

Vol. 61: Reports of the Midwest Category Seminar II. IV, 91 pages. 1968. DM 9,60 / $ 2.40

Vol. 62: Harish-Chandra, Automorphic Forms on Semisimple Lie Groups X, 138 pages. 1968. DM 14,– / $ 3.50

Vol. 63: F. Albrecht, Topics in Control Theory. IV, 65 pages. 1968. DM 6,80 / $ 1.70

Vol. 64: H. Berens, Interpolationsmethoden zur Behandlung von Approximationsprozessen auf Banachräumen. VI, 90 Seiten. 1968. DM 8,– / $ 2.00

Vol. 65: D. Kölzow, Differentiation von Maßen. XII, 102 Seiten. 1968. DM 8,– / $ 2.00

Vol. 66: D. Ferus, Totale Absolutkrümmung in Differentialgeometrie und -topologie. VI, 85 Seiten. 1968. DM 8,– / $ 2.00

Vol. 67: F. Kamber and P. Tondeur, Flat Manifolds. IV, 53 pages. 1968. DM 5,80 / $ 1.45

Vol. 68: N. Boboc et P. Mustatǎ, Espaces harmoniques associés aux opérateurs différentiels linéaires du second ordre de type elliptique. VI, 95 pages. 1968. DM 8,60 / $ 2.15

Vol. 69: Seminar über Potentialtheorie. Herausgegeben von H. Bauer. VI, 180 Seiten. 1968. DM 14,80 / $ 3.70

Vol. 70: Proceedings of the Summer School in Logic. Edited by M. H. Löb. IV, 331 pages. 1968. DM 20,– / $ 5.00

Vol. 71: Séminaire Pierre Lelong (Analyse), Année 1967-1968. VI, 19 pages. 1968. DM 14,– / $ 3.50

Vol. 72: The Syntax and Semantics of Infinitary Languages. Edited by J. Barwise. IV, 268 pages. 1968. DM 18,– / $ 4.50

Vol. 73: P. E. Conner, Lectures on the Action of a Finite Group. IV, 123 pages. 1968. DM 10,– / $ 2.50

Vol. 74: A. Fröhlich, Formal Groups. IV, 140 pages. 1968. DM 12,– / $ 3.00

Vol. 75: G. Lumer, Algèbres de fonctions et espaces de Hardy. VI, 80 pages. 1968. DM 8 – / $ 2.00

Vol. 76: R. G. Swan, Algebraic K-Theory. IV, 262 pages. 1968. DM 18,– / $ 4.50

Vol. 77: P.-A. Meyer, Processus de Markov: la frontière de Martin. IV, 123 pages. 1968. DM 10,– / $ 2.50

Vol. 78: H. Herrlich, Topologische Reflexionen und Coreflexionen. XVI, 166 Seiten. 1968. DM 12,- / $ 3.00

Vol. 79: A. Grothendieck, Catégories Cofibrées Additives et Complexe Cotangent Relatif. IV, 167 pages. 1968. DM 12,– / $ 3.00

Vol. 80: Seminar on Triples and Categorical Homology Theory. Edited by B. Eckmann IV, 398 pages. 1969. DM 20,– / $ 5.00

Vol. 81: J.-P. Eckmann et M. Guenin, Méthodes Algébriques en Mécanique Statistique. VI, 131 pages. 1969. DM 12,– / $ 3.00

Vol. 82: J. Wloka, Grundräume und verallgemeinerte Funktionen VIII, 131 Seiten. 1969. DM 12,– / $ 3.00

Vol. 83: O. Zariski, An Introduction to the Theory of Algebraic Surfaces. IV, 100 pages. 1969. DM 8,– / $ 2.00

Vol. 84: H. Lüneburg, Transitive Erweiterungen endlicher Permutationsgruppen. IV, 119 Seiten. 1969. DM 10,– / $ 2.50

Vol. 85: P. Cartier et D. Foata, Problèmes combinatoires de commutation et réarrangements. IV, 88 pages. 1969. DM 8,–/$ 2.00

Vol. 86: Category Theory, Homology Theory and their Applications I. Edited by P. Hilton. VI, 216 pages. 1969. DM 16,–/$ 4.00

Vol. 87: M. Tierney, Categorical Constructions in Stable Homotopy Theory. IV, 65 pages. 1969. DM 6,–/$ 1.50

Vol. 88: Séminaire de Probabilités III. IV. 229 pages. 1969. DM 18,–/$ 4.50

Vol. 89: Probability and Information Theory. Edited by M. Behara, K. Krickeberg and J. Wolfowitz. IV, 256 pages. 1969. DM 18,–/$ 4.50

Vol. 90: N. P. Bhatia and O. Hajek, Local Semi-Dynamical Systems. II, 157 pages. 1969. DM 14,–/$ 3.50

Vol. 91: N. N. Janenko, Die Zwischenschrittmethode zur Lösung mehrdimensionaler Probleme der mathematischen Physik. VIII, 194 Seiten. 1969. DM 16,80 / $ 4.20